Finsler Set Theory: Platonism and Circularity

Translation of Paul Finsler's papers on set theory with introductory comments

Edited by David Booth
and Renatus Ziegler

Springer Basel AG

Editors' Addresses

David Booth
Three Fold Foundation 307
Hungry Hollow Road
Chestnut Ridge
New York 10977
USA

Renatus Ziegler
Mathematisch-Astronomische Sektion am Goetheanum
4143 Dornach
Switzerland

Mathematics Subject Classification (1991): 03A05, 03E05, 03E20, 03E30, 04A20

A CIP catalogue record for this book is available from the Library of Congress, Washington D.C., USA

Deutsche Bibliothek Cataloging-in-Publication Data

Finsler set theory: Platonism and circularity : translation of
Paul Finsler's papers on set theory with introductory comments
/ ed. by David Booth and Renatus Ziegler. - Basel ; Boston ;
Berlin : Birkhäuser, 1996
   ISBN 978-3-0348-9876-8   ISBN 978-3-0348-9031-1 (eBook)
   DOI 10.1007/978-3-0348-9031-1
NE: B ooth, David [Hrsg.]; Finsler, Paul

© 1996 Springer Basel AG
Originally published by Birkhäuser Verlag in 1996
Softcover reprint of the hardcover 1st edition 1996
Printed on acid-free paper produced from chlorine-free pulp. TCF ∞
Cover design: Markus Etterich, Basel
Cover illustration: Photo of Paul Finsler by courtesy of Regula Lips-Finsler, handwritten manuscript of Paul Finsler by courtesy of the Mathematical Institute, University at Zürich

ISBN 978-3-0348-9876-8

9 8 7 6 5 4 3 2 1

Contents

vi

Foreword

Paul Finsler's set theory is the least explored of all the set theories which appeared after the noise and upheaval that followed in the wake of Russell's paradox. It remains a storehouse of relatively unexplored lines of thought, one of which, the existence of non-well-founded sets, has recently attracted new attention.

Paul Finsler (1894-1970) had a secure reputation as a differential geometer when he entered the tumultuous debates over the foundations of mathematics. But, if we imagine him simply as a non-specialist commentator on foundational controversies, it would hinder one's grasp of his theory. It is better to think of him as the heir to the spirit of Georg Cantor's set theory. Not only did they share an uncompromisingly Platonist philosophy but in other respects too Finsler carried on the outlook of Cantor; the distinction between sets and classes, for instance, which was developed by Finsler, first appears embryonically in Cantor's writings.

In spite of the fact that many of the papers here are quite old, they will also seem fresh and unusual. They contain the thoughts of a tragically solitary figure who pondered them intensely and they cause one to think about fundamental ideas that may have long been taken for granted.

In entering into the theory one may think of it as an attempt to recover the lost paradise of Cantor, the absolute universe of mathematics. This is not so much a world of infinite sets, ordinal numbers, or topological spaces, though these things all appear within it. The real universe is that of pure concepts. That we learn to recognize error, that we feel that science is capable of progress, all these things are evidence that there is a logical structure to concepts which stands above all logical calculi.

The *philosophical section* of this book contains the most important of Finsler's attempts to develop these ideas. His anticipation of Gödel's theorem in [1926a], for example, was an attempt to show that logic transcends all formal, deductive systems. As a steadfast Platonist in a period dominated by formalist and nominalistic ideas, Finsler embroiled himself in controversies now and then. These are of considerable historical interest, but would make little sense to a reader who was not already acquainted with the constructive works translated here.

The *foundational section* contains Finsler's set theory. We believe that most of the major arguments made in the theory have been included. The theory can be decomposed into the big Finsler theory and the little Finsler theory. The little theory involves a graphical, combinatorial attitude toward sets. The non-well-founded sets are its most characteristic feature. The big theory is more closely related to Finsler's philosophical view: The startling distinction between circular and non-circular sets is central to this part of his system.

What we have called the *combinatorial section* is an aspect of the little theory that can be separated off independently. Believing that sets could be viewed as generalized numbers, Finsler introduced number theoretic ideas into the study of finite sets. We have omitted his exploration of the countable ordinals, Finsler [1951]; and we have included an introduction to the combinatorics of finite, non-well-founded sets.

The papers of the philosophical and foundational parts are closely interwoven. One might misinterpret Finsler set theory were one quite unacquainted with his philosophical perspective. According to Finsler, the axioms of set theory rest on the actual interconnection of concepts not on formal, linguistic foundations. So it is potentially misleading to approach them as though they were merely crude principles in dire need of formalization.

The bibliography lists Finsler's pertinent writings, citations of them, references made within them, and some closely related studies. We have attempted to be complete within these boundries, but it is probably wishful thinking to suppose that nothing was omitted. We did not attempt to include a complete bibliography on non-well founded sets, axioms of set identity, principles of completeness in set theory, nor expositions of Platonism in mathematics. The bibliography of Aczel [1988] contains some references concerning relative consistency and formal systems that we have left out; in addition, his bibliography covers non-well-founded sets extensively.

Bibliographic citations refer to the publication date and, if appropriate, to the page number. Thus Bernays [1922, 15] refers to page 15 of a 1922 paper of Bernays listed in the bibliography. Finsler's papers which are included in this collection are marked within the bibliography with a star.

Our translation is deliberately not literal. It reflects our understanding of Finsler's ideas. We tried to make Finsler's intentions as clear as possible to professional mathematicians and philosophers. Historians and philologists might sometimes resent the freedom we took in rendering Finsler's arguments into understandable English. However, if someone wants, for historical or other reasons, to understand *all* the subtle details and references in Finsler's writings, he or she has to go back to the German original anyway. It should be mentioned that Finsler's marginal notes which were included as footnotes in Finsler [1975] have been incorporated directly into the text of these translations.

We are pleased to acknowledge the great help of David Renshaw of the University of Edinburgh, who made the preliminary version of these translations.

Georg Unger of the *Mathematisch-Physikalisches Institut* of Dornach, Switzerland, kept the Finsler theory alive, patiently and faithfully sought to preserve the theory in a form that is true to its intended meaning, and consulted with us on many aspects of Finsler's ideas. The collection of articles edited by Unger, Finsler [1975], includes some papers that have not been translated here. It also contains a valuable introduction to some aspects of Finsler's philosophical ideas.

With the support of

Science and Mathematics Association for Research and Teaching
Chestnut Ridge, New York, U.S.A.

Green Meadow Waldorf School, Chestnut Ridge, U.S.A.

Mathematisch-Astronomische Sektion am Goetheanum
Dornach, Schweiz

Mathematisch-Physikalisches Institut, Dornach, Schweiz

I. Philosophical Part

Introduction

Finsler's attitude towards mathematics was Platonistic in a very definite sense: He believed in the reality of pure concepts. Together they form the purely conceptual realm which encompasses all mathematical objects, structures and patterns. This realm exists independently of any particular state of human consciousness or individual experience. Mathematicians do not invent or construct their structures and propositions; they recognize, or discover, how these objects in the conceptual realm are interrelated with each other.

It is clear that if there exists a conceptual realm, then it must be absolutely consistent; hence *existence implies consistency*. This implication, however, does not suffice to prove the existence of pure concepts. The *Platonistic perspective* of mathematics can be expressed by the converse implication: *Consistency implies existence*. If a concept has been found to be consistent, it can be assumed to exist. This means that one can find properties and prove theorems about it.

At first, it may seem unnecessary to ask whether a conceptual entity is real, or has an ontological status beyond its consistency. Exactly this question, however, was at stake during the foundational crisis. Many critical thinkers contended that it is precisely this naive notion of existence that lies at the heart of foundational problems. Lacking a consistent and convincing Platonist philosophy, Hilbert, and many other mathematicians and logicians along with him, required that a mathematical object must be expressible in some language to really exist. Hilbert's approach to foundations tied mathematical existence to symbolic representation, that is to linguistic expressions in a strictly formal language.

Finsler entered the debate at this very point. He maintained that consistency is sufficient for the existence of mathematical objects. Furthermore, he thought that the antinomies which led to the foundational crisis, could be solved without the notion that existence is equivalent to formal constructibility. His main intention, visible in his writings, was to go back to the very roots of a strictly Platonistic interpretation of mathematics as in Cantor's set theory. Hence, Finsler's thoughts require a re-examination of basic issues in the philosophy of mathematics that are still unsolved, or at least have solutions that are not universally accepted.

Contrary to Cantor, however, Finsler never discussed his philosophical perspectives at any length. He assumed them to be self-explanatory for working mathematicians, that is, he assumed them to be clear from their experience. This might have been the case for most mathematicians in the 19th century but certainly not for the critical way of thinking which emerged from the foundational crisis.

In this introduction there will be a short reconstruction of Finsler's philosophy of mathematics. This is still a largely unexplored territory with many open problems. However, it is indispensable for an understanding of his purely mathematical research. It can be shown that Finsler's point of view is at the very least internally consistent, that is, an hypothesis which has to be taken seriously. Apart from that, it is inspiring and may start some fruitful future research.

One of the most important distinctions for a Platonist outlook on mathematics is the one between pure concepts and their verbal or symbolic representation. The latter is no substitute for the former: An expression merely points to the structure or pattern that it refers to. The pure concept is accessible to someone who makes the effort to *think* what is meant by such a linguistic expression. In particular, the notions of truth and consistency have their primary meaning beyond language: Their structure is, in the first place, not a matter of linguistic distinctions (for example between object language and metalanguage) but of *understanding* or *insight*.

It may be an easy matter to change the notation in which a theorem or a definition is expressed. We often translate – mathematicians are well accustomed to it – a theorem into some other language. The theorem itself, however, to which these notations or formulations refer, is invariant under merely linguistic transformations. The theorem itself cannot be altered. It is something which we become distinctly aware of as soon as we really *think* about it.

The realm of pure concepts is accessible by *insight*, or pure thought (some mathematicians – for example Gödel [1964] – call this "mathematical intuition"). This is part of the everyday experience of a mathematician, not something mystical. He might prefer to call it *informal thinking*, or more appropriately *nonformal*, instead. He experiences it above all in those moments in which he is not simply manipulating symbols or in which his thoughts have not yet been symbolically expressed. In particular, logical calculi, like all calculi, are manipulative (done by hand or machine) so they cannot capture thinking, though they may reflect it.

What is meant by "informal thinking"? There is a whole range of conceptual qualities to which this might refer: From a well thought out theory to a very vague, or even highly speculative, conjecture. These individually conceived conceptual structures all have in common this *nonformal* nature.

When a mathematician has an idea which gives a new insight and is important for his topic of research, then he tries, at least in principle, to organize his thoughts into a rigorous, deductive pattern of arguments which represents the original nonformal insight as closely as possible. This process, although it might include writing or symbol manipulation, is conceptual in its essence. Even if one goes as far as expressing one's ideas within a strictly formal language, the main goal is still to represent the initial idea adequately. Such procedures are reasonable ones; it may appear, however, as if the clarity and rigor of the final structure is due to its formal style of presentation. But how was the clarity and security of the intended informal thought patterns achieved? What are the criteria by which a mathematician judges the process and final result of the formal in comparison with the preformal stage? It is by his or her own nonformal insight, or understanding, which started and accompanied the whole process of formalization. This shows that the nonformal insight is prior, systematically and temporally speaking, to the formal one. We would not know what to formalize if it were not for the nonformal insight, the pure concept that we are aware of. *Formalization occurs at the end and not at the beginning of the true path to mathematical understanding.*

An opponent of Platonism might argue that preformal insight is vague by its very nature and hence cannot really be the source for any precise mathematics. Platonists, however, do not argue against organizing the initial fuzzy thoughts or intuitions into rigorous chains of arguments based on a set of clearly defined assumptions or axioms. Furthermore, they are aware of the fact that in writing something down, one increasingly clarifies the ideas. But, independently of how far one goes in spelling out the details symbolically, it is still the nonformal insight which guides writing and not the syntactic rules of language.

Consequently, the final linguistic expression is a mere *representation* of the real thought – and is not to be confused with this thought, or concept, itself. If a symbolic expression is given, one usually refers to the corresponding thought or concept as its *meaning* or *content*.

Let us come back to our primary distinction between conceptual content and linguistic expression. From this point of view, the distinction between *object language* and *metalanguage* (or mathematics and metamathematics) presents itself as a projection of the former distinction onto the realm of language. Without the former distinction, the latter would be artificial. This becomes evident if one proceeds to formalize the metalanguage itself. In this case, assertions of truth, meaning etc. about an expression in the given formal language (object language) appear just as another string of symbols which in itself do not explain themselves but would need a meta-metalanguage. In practice one usually uses natural language as a metalanguage (including the meta-metalanguage etc.) of the given object language. However, natural language is only another means of expression which bears neither truth nor meaning in itself but asks for some non-linguistic, i.e. conceptual interpretation. Effectively, truth and meaning can only be found in the pure conceptual realm if one does not want to fall into an infinite regress, that is, an open-ended hierarchy of languages.

When a Platonist like Finsler refers to a theory, to mathematical objects, or to a set of axioms, he refers to objects in the purely conceptual realm. The specification of a formal *language* has no part in his purely *mathematical* deliberations. *Mathematics is concerned with relationships between concepts and not with the expression of concepts in language.*

Of course, a Platonist also represents his thoughts with the help of languages, but he is well aware of the fact that it is his insight which gives meaning to his words and not the other way round.

What then, one might ask a Platonist, is the function of language? Why use it at all?

The primary purposes of language in mathematics are *communication*, *symbolic computation*, and *checking*. As far as mathematical *insight* goes, there is, strictly speaking, no need for a language. *Mathematics is not the science of communication of structures and patterns, but the science of these structures and patterns themselves.* However, if one wants to tell someone else about one's discoveries, there is no way around using some kind of language to express them. In addition, it is helpful for storing one's thoughts (in the form of their symbolic representations) in an external memory, or for checking the results by some well-known computational methods.

As for *symbolic computation*, the need of appropriate notations for accurate and efficient symbol manipulations is evident. However, the meaning of the symbols and the rules of computation do not emerge merely from the rules of the syntax nor the grammar of the relevant language. Calculations are based upon a set of rules that are implemented in language from a realm outside of it. Thus, the results of computation need to be interpreted, apart from merely symbolic checks.

However, if it were not for communication or symbolic computation, there would be no necessity for language; mathematical insight would still be there without any language. Mathematicians might write down their ideas or compute something symbolically in order to *check* the results against some prior knowledge or with acknowledged, secure methods. However, they do not need to write their thoughts down in order to *understand* them. If this were the case in its strictest sense, how could they ever know the meaning of what they wrote down?

We need to be careful not to confuse the complicated and sometimes rather "irrational" search for conceptual clarity, during which we might go through different stages of computing, writing and editing, and the purity of insight that we arrive at in the end. We are only concerned here with the latter: The final clear insight. It transcends the symbolic patterns, as everyone knows who tries to write or read *and* understand a mathematical paper; it is not enough to recognize symbols, to know their syntactical structure, or to be able to follow the pattern of a symbolic computation. One needs to think and thus grasp the meaning of the thoughts which are to be communicated.

From this point of view it should be clear that most Platonists are not interested in the fine structure of a language for its own sake, but only as a means of expressing pure thoughts. What they want to understand are the concepts themselves, not just verbal or symbolic representations.

Let us now turn to Finsler's philosophical papers from the standpoint of an historian, taking leave of the Platonist point of view.

Are there Contradictions in Mathematics? [1925]

This paper is a preview of Finsler's future research on the foundations of mathematics, set theory in particular. To begin with, he announces his intention to restore the consistency of mathematics by solving, not avoiding, the antinomies. One does not need a new logic nor a correction of the old one for this purpose.

In dealing with what Finsler calls "logical antinomies" (later called "semantical antinomies"), i.e. the antinomies of the "liar" and the antinomy of finite definability, he introduces the distinction between the explicit and the implicit content of a proposition. The "explicit content" refers to the conceptual meaning and the "implicit content" to the form of representation. Antinomies arise if these two "contents" contradict each other.

Concerning the "set theoretic antinomies", in particular Russell's paradox, Finsler points out that one needs to distinguish between satisfiable and unsatisfiable circular definitions. Russell's definition of the set of all sets which do not contain themselves is a non-satisfiable circular definition. Finsler, however, maintains that it is not necessary to exclude all circular definitions because of that; they are used even in algebraic equations.

Solving the antinomies does not positively solve the problem of a consistent foundation of set theory. That task is reserved for the paper *On the Foundations of Set Theory* [1926b] in Part II of this book.

Formal Proofs and Decidability [1926a]

In this paper Finsler establishes the formal undecidability of a proposition which is, however, false. From this he concludes that formal consistency does not imply absolute consistency.

In carrying out his proof, Finsler is not so much concerned with a precise definition of a formal system as with the demonstration of the limitation of any kind of symbolic representation. In order to show that there *are* formally undecidable propositions, he refers to the fact that any language uses at most countably many symbols. Hence, not all propositions of the form:

α is a transcendental number,

are expressible, or definable in a language, since there are uncountably many transcendental numbers. Consequently, since only countably many of these propositions can be formally proved within the given language, there must exist propositions of this kind which cannot be proved in these terms but which are still true.

Finsler goes on to present an example of a proposition that is formally undecidable yet false. In order to show the latter, he refers to the conceptual content of the verbal expression in question. He shows that if this conceptual content is taken into account, then the formally undecidable proposition turns out to be false.

One might summarize the argument here as follows: If there is a purely conceptual realm, no formal representation can capture it.

In effect, Finsler's main intention is not to distinguish between different kinds of formal systems but between the purely conceptual realm and its symbolic ("formal") representation, including the use of natural language. This is why he did not need to specify more precisely his notions of formal proofs, formal definability, formal systems etc.; every thing which is written down is formal in Finsler's sense. Hence, for his purpose, there is no need of a general reconstruction of language.

From this, the comparison between Finsler's incompleteness argument and Gödel's incompleteness proof [1931] takes on a new perspective. There are indeed striking similarities between Finsler's and Gödel's approach. However, as Van Heijenoort [1967] remarks in his introduction to Finsler's paper,

> Finsler's conception of formal provability is so profoundly different from Gödel's that the affinity between the two papers should not be exaggerated. [1967, 438]

This is certainly true, since Gödel's most profound achievements lie in the accurate definition of the particular formal systems in question and the concept of a formal proof within this system. Furthermore, he developed what is now called "the arithmetization of metamathematics"; for this purpose he gave a precise definition of the class of recursive functions. By precisely defining his formal methods, he shows, by constructing an example of an undecidable statement, that these formal methods are incomplete. An additional metamathematical argument then shows that this proposition which states its own unprovability is, in fact, true, and hence decidable on the metamathematical level. By these means Gödel achieved something that Finsler had not done: He proved even for the strictest formalist that formal means have their limits (see Dawson [1984] for further elaborations on this point).

Solely from the point of view of mathematics and formal logic, Gödel's paper is far more significant than Finsler's. However, Finsler's paper goes directly to the heart of the philosophical problem. Finsler is concerned with the fundamental distinction between concepts and their symbolic or verbal representation, not with the formally more sophisticated but philosophically limited distinction between metalanguage and object language. One might say that the latter distinction is the projection of the former distinction onto the realm of language.

Gödel was acutely aware of the objective Platonist principles behind the distinction between mathematics and metamathematics. Later in his life, he expressed strongly Platonist convictions, for example in the essays *Russell's mathematical logic* [1944] and *What is Cantor's continuum hypothesis* [1964], although he never exhibited these in his earlier writings on the foundations of logic. Fefermann [1988] argues that Gödel's extreme caution towards the power of formalist views of his time urged him to shy away from expressing his Platonist convictions until the Forties.

The audience Gödel wanted to address consisted of strict formalists. Their opinions were the only ones that mattered to him. This is why he restricted his analysis to the concepts and methods *they* could accept, namely, semantic distinctions, syntactic forms, restrictions to particular formal systems, and relative rather than absolute consistency. From the perspective of the strict formalist, apparently, what Finsler has done is "nonsensical" (see the quotation of Gödel in Dawson [1984, 82f]), since it presupposes something the formalists reject: the existence of the purely conceptual realm. For instance, J.C. Webb thinks that the main achievement of the mechanization of Finsler's argument by Gödel was to bring "Finsler's undecidable sentences down from [the] "rein Gedankliche[n]" and put them back into the formal system. In short, he formalized Finsler's diagonal argument." [1980, 193]. Hence, in his opinion, there is no threat to mechanist or finitist convictions any more: There is nothing left a machine could *not* do. Webb misses the point, predictably however for a formalist of his persuasion, that this projection is not possible without severe philosophical effects, as was shown above.

It is not appropriate, however, to judge Finsler from this formalist point of view, even though he himself sometimes thought so (cf. Dawson [1984, 81]). Finsler wanted to prove that Platonism is a consistent and fruitful philosophical perspective (cf. Finsler [1941a]), by developing foundations for set theory in [1926b]. As a consequence of the arguments in his paper on *Formal Proofs and Decidability*, he could not accept any kind of formal restrictions concerning set theory, because set theory lies at the very heart of the *foundations* of mathematics itself. No formalized theory can ever capture foundational conceptions that bear upon *all* of mathematics. To use geometrical terms, formal theories apply only to local structures, not to global ones.

In concluding, it is important to note that no strict formalist will ever be convinced by Finsler's paper on *Formal Proofs and Decidability* [1926a], because Finsler assumes something that formalists cannot accept: the reality of pure concepts. Finsler did not make clear what he meant by that; this certainly limits the significance of his paper. However, we cannot exclude the possibility that the open questions about the consistency of the Platonist perspective of mathematics and the ontological status of the realm of pure concepts may be solved some day. Even Gödel [1964] could not say more than Finsler concerning his belief in the objective existence of the objects of mathematical intuition. Gödel chose not to refer explicitly to the reality of concepts in his purely mathematical research, whereas Finsler boldly did so.

On the Solution of Paradoxes [1927b]

In this paper Finsler expands on his ideas in the paper [1925] concerning the solution of paradoxes which involve circular definitions.

Are there Undecidable Statements? [1944]

Here Finsler compares his approach to incompleteness with Gödel's. His arguments are closely related to the "liar". He begins with a discussion of this paradox. What then follows is one of Finsler's most original contributions to the analysis of the semantic paradoxes. Finsler shows that there is an absolutely consistent proposition and that there is a statement which an individual mind cannot prove yet has to believe.

Some choose to call this paper "obvious nonsense" or even "almost pathological" without further elaborations (see Dawson [1984, 83]). We hope that this translation makes Finsler's arguments more accessible and less subject to misunderstandings.

The paper begins with the fundamental distinction between "formal" and "inhaltlich"; this is the distinction between formal representation within a symbolic language (or linguistic expression in general) and conceptual content. This distinction is instrumental in proving that there are, in principle, undecidable statements relative to a particular formal system which are nevertheless decidable conceptually, that is, decidable in an absolute sense.

Finsler maintains that it might superficially appear that Gödel referred to the conceptual realm when he showed, through a metamathematical argument, that there is a statement unprovable within the formal system which is decidable in the metasystem. However, if one takes into account that the metasystem can also be formalized, Gödel's incompleteness result only shows that there are undecidable statements *relative* to a given formal system. Such a system can always be enlarged in order to make the statement in question decidable. But then there will be another undecidable statement in this larger system and so on.

In view of this, Finsler argues that Gödel did not prove the existence of a proposition which is formally undecidable in principle. Finsler maintains that if one strictly *requires*, as a matter of principle, the formalization of all arguments involved (including the metamathematical ones), then Gödel's result becomes contradictory: The formally undecidable statement becomes formally decidable. This contradiction only disappears if one explicitly takes into account the conceptual realm, or if one severely restricts the available logical tools on the object level.

One might object that Finsler's arguments are only correct if one ignores the distinction between the object level and the metalevel. Indeed, this distinction is one of the major achievements of modern mathematical logic. However, Finsler never refers to it. Was he not aware of it? Or did he simply ignore it?

In fact, this distinction is of minor importance within Finsler's approach. He was not interested in consistency, completeness, decidability, etc. *relative* to a certain formal system, but in *absolute* consistency, in short, in *absolute* results. Consequently, he was not interested in studying the subtle effects of modifications, restrictions, or extensions of various formal systems, but in the analysis of the effects of formal representation itself. Hence there was no need for him to distinguish between the object language and the meta-language. This distinction exists only for concepts that are expressed in language.

Gödel's unpublished remark that Finsler's aim, to achieve absolute results, is "nonsensical" was at the very least hasty. After all, Gödel himself refered from time to time to absolute notions (see Dawson [1984] and Fefermann [1988]).

This paper on absolute decidability shows clearly what Finsler wanted: that mathematical thinking not artificially limit itself by requiring that formalization be an *essential* part of mathematical existence.

The discussion of the "liar" in §2 is based upon the distinction between the explicit conceptual content of a proposition and its implicit assertion that it be true or false. The paradox arises out of the fact that the implicit assertion which contradicts the explicit assertion, is ignored.

In §3 Finsler expands the notion of proof so that it includes all possible ideal proofs. He can then show that the assumption that there are no undecidable statements (i.e. no unsolvable mathematical problems) is absolutely consistent. In particular, he shows that it is impossible to prove that a certain proposition is absolutely undecidable. From this he deduces, in §4, one of his most original results: There is a statement, which I, myself, cannot prove yet need to believe, because it can be proved rigorously by someone else.

The Platonistic Standpoint in Mathematics [1956a]

This paper records Finsler's part of a discussion of foundational issues in the journal *Dialectica*. It contains a reference to Specker's objection which is treated in sections VII and VIII of the introduction to the *Foundational Part* of this book.

Platonism After All [1956b]

　　As in the paper above, this is Finsler's part of a discussion, not all of which is included here (see Wittenberg [1956], Bernays [1956], Lorenzen [1956]). It contains mention of Ackermann's set theory [1956], which is described in section X of the introduction to the *Foundational Part* of this book.

14

Intrinsic Analysis
of Antinomies and Self-Reference

Introduction

This essay proposes a reconstruction of the ideas lying behind Finsler's analysis of antinomical situations. We do not think that it is necessary to give a literal account of his arguments: They can be easily followed in his papers. Instead we sought a basis that the various arguments might have in common. This essay is self-contained. It does not depend on the *results* of Finsler's analysis but gives an independent account of antinomical and self-referential situations.

Since Finsler's first philosophical paper [1925], antinomies have been studied in many different ways. Most authors believe that they require us to make some revisions either in our language or in our ways of thinking (cf. for example Quine [1962]). It is generally agreed that there has been no satisfactory solution to the antinomies; those accounts which take place in a formal system seem too restricted to provide a *complete* analysis.

Formal systems are not used in this paper. As Finsler argued so forcefully, formal systems might not possess the flexibility necessary to deal with the antinomies. We do not want to construct an artificial language in order to avoid the antinomies or, as has been tried recently by Barwise/Etchemendy [1987] to incorporate the antinomies into a formal structure: It is ultimately necessary to diagnose the antinomies on their own ground. We should make a self-contained, or *intrinsic analysis* which does not introduce assumptions foreign to the problem but focuses on the realm from which the antinomies arise. In particular, no new conceptions of truth nor theories about the limitation of human thought are needed. We shall use common logic, the traditional distinction between symbols and their meaning, as well as the distinction between conceptual and perceptual facts.

Summary

Antinomies involve perceptually or conceptually distinct objects which become identified during a line of thought, thus producing a contradiction. The intrinsic analysis keeps the distinct objects separate while allowing one to see how they can become identified. The analysis of antinomies calls for distinctions that are present in self-referential situations in general. They are easily overlooked, however, if no contradiction arises. – Semantic and logical antinomies have a similar pattern, though they operate in different realms. Their emergence gives insight into the formation and ontological status of concepts.

Part I: Antinomies

1. On the distinction between semantic and logical antinomies

The distinction between semantic and logical antinomies goes back to Ramsey [1926]. *Semantic* antinomies involve explicit reference to symbols or sentences, that is, both linguistic expressions and their meanings are present. *Logical* antinomies, however, involve only concepts and conceptual relations from mathematics and logic.

If we formalize the semantics as in model theory, the real distinction between the semantic and logical antinomies is lost. We will not treat these antinomies formally here. In effect, we claim that mathematical logic does not contribute to the *solution* of the antinomies. This does not mean that mathematical logic has not been enormously fructified through the analysis of the antinomies. It is to say, however, that there can be no *comprehensive* account within formal logic alone of how and why the antinomies arise.

An important result of our analysis is, however, that from a certain point of view, the structure of semantic and logical antinomies turn out to be the same. This point of view varies a great deal from mathematical logic. It is grounded in a detailed analysis of self-reference which lies at the heart of the antinomies.

2. Antinomies, contradictions, distinctions

By an *antinomy* we mean an argument which convincingly leads to a contradiction. By a *contradiction* we mean the conjunction of two mutually negated propositions. One proposition asserts that x has the property E and the other proposition asserts that x does not possess the property E :

(C) $(x$ is $E)$ and $(x$ is non-$E)$.

Surely we can *write* down such contradictions, as for example:

(12 is divisible by 3) and (12 is not divisible by 3);

but there has not been found any actual objects for which there is a property E such that (C) holds. Of course, the number 12 is no such object.

Evidently, to claim, that a house having red and green spots is both red and non-red does not constitute a contradiction in the sense defined above.

An *antinomy*, in its strictest sense, is not an argument that leads just to the conjunction of two mutually negated assertions, but it establishes the *equivalence* of two such opposite assertions. Therefore, an antinomy establishes the conjunction of two converse implications:

$$(x \text{ is } E \to x \text{ is non-}E) \text{ and } (x \text{ is non-}E \to x \text{ is } E).$$

In contrast to these notions, *distinctions* arise from observation. In particular, two objects x and x' are called *distinguishable* in case they are distinct and for some property E the following holds:

$$(x \text{ is } E) \text{ and } (x' \text{ is non-}E).$$

For example, circles and polygons are distinguishable objects by virtue of the latter possessing corners.

In some German philosophical literature (cf. for example Hösle [1986], Wandschneider [1993]) the terms "analytischer Widerspruch" and "pragmatischer Widerspruch" are used for contradictions and distinctions respectively. We shall follow this tradition and introduce the terms "analytical contradiction" and "pragmatic contradiction" for contradictions or distinctions respectively. If "contradiction" has no qualifier, then it stands for "analytical contradiction".

3. Derivation of a semantic antinomy

Let us turn now to a specific semantic antinomy. The following antinomy is a slightly modified version of Carnap's [1934] "liar cycle."

(1) a: b is true.
                                         b: a is not true.

The traditional argument goes like this: In case a is true, then b holds.   From this it follows that a fails – contradicting the assumption. On the other hand, in case *a* is false, then *b* fails. This makes *a* true – another contradiction. Thus we have an antinomy, the equivalence of two mutually negated assertions. Without doubt, the conclusion of this argument is a contradiction. An intrinsic diagnosis must by its very nature take hold of the course of the argument itself; it must reveal the root cause of the contradiction, not just a way to avoid it.

4. Diagnosis of a semantic antinomy

The argument deriving a contradiction in Carnap's Liar cycle begins with the first line of (1). There is a proposition *a*, which involves a sentence *b* in the second line; *b* in turn mentions *a*. For the derivation of the contradiction it is necessary that the *a* in the first line is identified with the *a* in the second line. *Without this identification there would be no contradiction.* Now, is this identification proper? The only essential property of *a* present in the second line is that *a* is the subject of a proposition. In the first line *a* does not stand for the subject of the proposition, but represents the whole proposition. Thus, the two *a*'s have a different meaning. Hence the two objects, called "*a*", in the first and the second line of (1) serve a different purpose and need to be distinguished clearly. Since for *b* we may argue along similar lines, we are presented with a new version of (1) that makes this distinction explicit:

(1') $a^{(1)}$: $b^{(1)}$ is true.
      $b^{(2)}$: $a^{(2)}$ is not true.

By strictly following the *principle of identity*, which says that only objects possessing identical properties may be identified, the contradiction evaporates. Only when $a^{(1)} = a^{(2)}$ and $b^{(1)} = b^{(2)}$ are assumed, ignoring the proper distinctions, can a contradiction result.[1]

The common form of the Liar is: "This sentence is not true." Here "this sentence" is the subject of the proposition "This sentence is not true" *and* refers to this proposition as a whole. If we abbreviate "this sentence" by "*s*" we can represent the full structure of "This sentence is not true" by:

(2)   *s*: *s* is not true.

---

[1] The essential features of this diagnosis where first pointed out to me by Werner A. Moser.

Here too we have to distinguish between the s before the colon and the s after the colon. The latter represents the subject of a proposition while the former stands for the proposition itself. Hence it is necessary to indicate the different meanings by different symbols as before:

(2') $s^{(2)}$: $s^{(1)}$ is not true.

Without identifiying $s^{(1)}$ and $s^{(2)}$ no contradiction results.[2]

5. *Some objections*

One could argue that $s^{(1)}$ designates not only the subject of the proposition $s^{(2)}$ but *also* this proposition itself. Hence $s^{(1)}$ and $s^{(2)}$ are the same. But in this case, the meaning of $s^{(1)}$ (as well as the meaning of $s^{(2)}$) would not be not unique any more. Namely $s^{(1)}$ has two mutually incompatible meanings: On the one hand $s^{(1)}$ is the subject of a proposition and stands on that account for a *part* of the proposition and therefore is distinguished from $s^{(2)}$. On the other hand, $s^{(1)}$ stands for the *whole* proposition and in this role is identical with $s^{(2)}$.

If we were to assume, in order to identify $s^{(1)}$ and $s^{(2)}$, that $s^{(1)}$ represents simultaneously the subject of the proposition *and* the whole proposition, then $s^{(2)}$ must also be given a new meaning. In this case, $s^{(2)}$ is supposed to represent not just a proposition with a subject and a predicate, as before, but also this subject itself. In other words, $s^{(2)}$ is a circular proposition. Obviously, this throws us back into the situation of (2), taking s to signify the proposition as well as the subject of this proposition.

A moment's thought shows that this interpretation is incomplete. We would have to take into account the new situation that arises from the identification of $s^{(1)}$ and $s^{(2)}$. Now, the s in front of the colon does not only represent the proposition and its subject, as does the s that follows the colon, but also a proposition about a proposition. This argument shows that we have to write:

---

[2] Goddard/Johnston [1983] argue along somewhat similar lines. They realize that in order to derive the antinomy, one has to assume the identity of two structurally distinct parts. Because they work within predicate calculus, the scope of their analysis is more limited than ours: an intrinsic analysis demands that the nature of the predicates be analyzed; this goes beyond predicate calculus. Hence, Goddard/Johnston cannot explain why and how but only *that* differentiations have to be made in order to derive a contradiction.

(3) $s^{(2)}$: $[s^{(1)}$: $s^{(1)}$ is not true].

But our analysis above shows that this representation is not sufficiently precise, because the two instances of $s^{(1)}$ are distinct and therefore cannot have the same meaning. When we apply the same procedure that was used with (2) to the expression within the brackets of (3), we arrive at

(3') $s^{(3)}$: $[s^{(2)}$: $s^{(1)}$ is not true].

As in (2'), this expression does not give rise to a contradiction, as long as we do not ignore the distinction between the increasingly numerous instances of s.

The transition from (2') to (3) arises from the desire to eliminate any distinction between $s^{(1)}$ and $s^{(2)}$. So we would have to give $s^{(1)}$ an additional property that naturally belongs to $s^{(2)}$ in order to blur the distinction between them. But this effort does not succeed. Since the main structure of (2) is represented within (3), we must substitute (3') for (3). Continuing this process we would have to give $s^{(2)}$ the properties of $s^{(3)}$. In turn this gives a new expression $s^{(4)}$ that has an additional property not possessed by $s^{(3)}$, and so on.

(3$_3$') $s^{(3)}$: $[s^{(2)}$: $s^{(1)}$ is not true].
(3$_4$') $s^{(4)}$: $[s^{(3)}$: $[s^{(2)}$: $s^{(1)}$ is not true]].

.

.

.

(3$_n$') $s^{(n.)}$: $[s^{(n.-1)}$: $[...[s^{(2)}$: $s^{(1)}$ is not true]...]].

The infinite regress never forces us to the the conclusion that there is not really any distinction between $s^{(k)}$ and $s^{(k+1)}$:   for any $k =$ 1, 2, 3, ... The only way that a contradiction can ever appear is for us to assume, contrary to the arguments given above, that

$$s^{(1)} = s^{(2)} = ... = s^{(n)} = ...$$

The contradiction cannot be constructed unless we ignore a factual distinction.

It is often argued that the expression

(2$_2$) s: s is not true;

lends itself to an iteration if we substitute for s the proposition: s is not true. This substitution gives the sequence that follows.

(2_3) s: [s is not true] is not true;

.
.
.

(2_n) s: [...[[s is not true] is not true]...] is not true.

In order to make the different steps of this iteration more explicit, we write:

(3_2) $s^{(2)}$: $s^{(1)}$ is not true.
(3_3) $s^{(3)}$: [$s^{(1)}$ is not true] is not true.
(3_4) $s^{(4)}$: [[$s^{(1)}$ is not true] is not true] is not true.

.
.
.

(3_n) $s^{(n)}$: [[...[[$s^{(1)}$ is not true] is not true]...] is not true] is not true

In iterating (2_2) it seems at first that we do not need to add any new property to this expression but just carry out what is already present. Hence (2_2) and (2_n) would be equivalent in spite of the fact that s in (2_2) has a simple subject and s in (2_n) has a much more complicated one. Obviously, this leads to a distinction, or "pragmatic contradiction".

We must observe, however, that the equivalence of (2_2) and (2_n) can only be established if we let $s^{(1)} = s^{(2)} = ... = s^{(n)}$ in (3_2) to (3_n), for all n. In fact, the different s's do not merely indicate different steps of the iteration but signify, according to our previous analysis, different meanings as well. This has already been shown for (3_2) in section 4.

Now consider the iteration itself. Each step is carried out by substituting $s^{(1)}$ for $s^{(2)}$. This definitely *changes the meanings* of the s in front of the colon with every step. In particular, the transition from (3_2) to (3_3) rests upon the fact that the s in front of the colon no longer signifies a proposition, but a proposition about a proposition.

What looks at first like an identification of $s^{(1)}$ with $s^{(2)}$ turns out to be the cause for a shift in the meanings of the s's in front of the colon. In fact, the different meanings of $s^{(1)}$ and $s^{(2)}$ are the driving forces of the iteration. Only by giving $s^{(1)}$ the meaning of $s^{(2)}$ can the iteration be started and kept going. Therefore, without taking into account these different meanings, there would be no iteration at all. We would be stuck with (2_2) and could never arrive at (2_n). If we deny the different meanings of the s's before and after the colon, then we have no complete argument to actually carry out the iteration. Denying these different meanings and producing the iteration are incompatible: They lead to inconsistent arguments.

The essence of the intrinsic analysis of the Liar antinomy consists in pointing out that distinguishable objects in the sense of section 2 are identified and hence have no identity as something individual. Not observing this difference is an offense against the principle of identity. Therefore, what is real are only the given distinct objects: The *antinomy* is not a fact of reality. The *antinomy* is *created* by the one who derives a contradiction while projecting distinguishable objects into the conceptual realm.

6. *Epistemological analysis of a semantic antinomy*

In this section it will be shown that a thorough analysis of the argument leading to a contradiction from the Liar cycle (1) includes epistemological categories. It will turn out that the *paradoxer*, namely the one who attempts to derive a contradiction, *actually* does differentiate between the different meanings of a and b in the first and the second line of (1). This differentiation is forced upon him. But later in the argument he chooses to ignore this distinction. The irony of the situation is that the paradoxer observes this distinction even while reasoning as though the two different uses of the parameters are identical. In particular, his observation requires the actual perception of concrete objects.[3]

(1) a: b is true.
                                           b: a is not true.

The assertion, a is true, i. e., [b is true] is true, can be applied to the second line of (1) if and only if we take it in the sense: a is *actually* true. Otherwise it would be an abstract proposition with no consequences for the b in the second line. Hence the a in the first line should be taken as an actual proposition about b: b *is* true. The concrete inspection of b in the second line shows that a is merely the subject of a proposition. The second line, standing alone, does not reveal that a is an *actual* proposition about something. As soon as the a in the first line and the a in the second line are *identified*, the

---

[3] According to Barwise/Etchemendy [1987] this means that the paradoxer has to take into account the situation the proposition is about, namely itself, taken as a concrete object of the world: a linguistic expression. Situations may include propositions, but are not propositions themselves. They are not purely conceptual but involve some kind of perception.

observed distinction is lost. (The same reasoning applies when a is taken to be not true.)[4]

An analysis which relates the first *and* the second line of the Liar cycle (1) has to take into account the different significance of the two instances of a, likewise for b. This different significance is, in effect, conceptual, but its conception involves a *perceptual* component. Our analysis cannot be purely conceptual in the sense that although the *result* might be purely conceptual, the object of our analysis is not.

The fundamental contrast appearing here is *not* that of two contradictory assertions but rather a pair of distinguishable objects in the sense of section 2. In effect, if the two distinguishable objects are projected into the purely conceptual realm, and if we also ignore the principle of identity, a contradiction does arise. More precisely, the a in the first line of the Liar cycle (1), called $a^{(1)}$, has the property, say E, not only to be a proposition, but to be an actual true proposition *about* the b in the second line. In short: $a^{(1)}$ is E. On the other hand, the a in the second line, called $a^{(2)}$, is merely the subject of a proposition not having a concrete significance. To sum it up: ($a^{(1)}$ is E) and ($a^{(2)}$ is non-E). This conjunction of two propositions has the form of a distinction in the sense of section 2. The identification of $a^{(1)}$ with $a^{(2)}$, namely the identification of the a with concrete signficance, called $a^{(1)}$, with the a with merely conceptual significance, called $a^{(2)}$, leads to a contradiction: (a is E) and (a is non-E).

We conclude that the antinomy, namely the argument leading to a contradiction, starts from a projection of two distinguishable objects (pragmatic contradiction) into the conceptual realm, that is, into the realm of logic. In the next step one neglects the distinction just made and thus ignores the principle of identity. The result is an analytic contradiction.

Let us discuss from this perspective Finsler's [1925] favorite example of a semantic antinomy. Consider a box with the following inscription:

1,   2,   3

The smallest natural number which is not specified in this box.

---

[4] Again, in the terms of the analysis of Barwise/Etchemendy [1987] the above argument amounts to the following: the paradoxer does in fact suppose a particular situation for "a" in the first line in that he applies this proposition to the second line. But later in his argument, the paradoxer acts as if he never made this supposition.

Which is this smallest number? There are only finite many numbers specified in the box. Hence there must be a smallest. Assume it is the number 4. But then 4 *is* specified in the box – hence it cannot be equal to 4. Conversely, assume this number is not equal to 4 but, say, equal to 5. It follows that 5 is *not* the smallest number which is *not* specified in the box. This gives a true antinomy: If the smallest number which is not specified in the box, call it x, is equal to 4, then it is not equal to 4. Conversely, if x is not equal to 4, then it is equal to 4.

Let "T" denote "specified on the blackboard" and let "a" denote "the smallest natural number >3". Then the propositions

$$b\colon a \text{ is not-}T,$$
$$a\colon b \text{ is } T,$$

have the same structure as the antinomy above. What is relevant for the existence of this smallest number >3 is the fact that it is written in the box, i. e. has the property T. Everything which is not explicitly written in the box is irrelevant. This means that the corresponding proposition is not true. Hence we conclude that this antinomy has the same structure as (1) if we substitute "T" by "true". Conseqently, the same method of analysis applies.

In short, the antinomy is produced by the pragmatic contradiction between the propositional content of "$b\colon a$ is not T" and the fact that b *is* written in the box. Identifying these two instances of b produces an analytical contradiction.

7. *A distinction without antinomical character*

The ideas introduced to analyze the Liar are of importance also for assertions without antinomical character but which nevertheless are problematic. Consider the assertion, sometimes called the truth-teller:

(4) $s\colon s$ is true.

No one finds a contradiction here: (4) is not an antinomy. And yet here too we must distinguish between the s in front of the colon and that which follows it. The latter, called $s^{(1)}$, is merely the subject of a proposition, and the former, called $s^{(2)}$, stands for a concrete proposition:

(4') $s^{(2)}\colon s^{(1)}$ is true.

In this case, the identification of $s^{(1)}$ with $s^{(2)}$ does not lead without further ado to a contradiction, but even so is as unjustified as in the antinomical form (2'). The previous discussion of antinomies requires ideas which are forced upon us by consideration of non-antinomical statements as well. The antinomies serve to introduce these ideas to our attention. It is unnecessary for us to develop strategies for escaping from antinomies. After all, the corresponding contradictions only arise out of our neglect of the above mentioned distinctions. The antinomical result deduced from the Liar situation is *our* creation, not something alien we have to defend against.

Clearly, $s^{(1)}$ and $s^{(2)}$ are distinguishable objects in the sense of section 2. By projecting $s^{(1)}$ and $s^{(2)}$ into the conceptual realm and simultaneously identifying them a contradiction arises, even with the truth-teller (4'). The truth-teller is, strictly speaking, not antinomical, but one still has to deal with distinguishable objects.

Part II: Self-reference

8. Self-referential assertions

The crucial problem presented to us by the antinomies is the *structure of self-reference.*[5]

The common feature of (2) and (4) that makes necessary the transition to (2') and (4') respectively is the self-referential structure of the corresponding assertions. Let E be any property. An *assertion* is called *self-referential*, if it has the following structure:

(S) s: s is E.

An essential *ingredient* of a self-referential *assertion* is a proposition whose subject is this proposition itself. A proposition standing alone cannot be self-referential.[6] It is essential here that part of the proposition, namely the subject, is associated with something outside the proposition. We are led beyond the proposition itself, understood as a mere conceptual entity, to the

---

[5] This point of view has also been put forward by Kesselring [1984: 104f.].

[6] This is to say that a proposition without any indication to its appropriate situation, or concrete significance, cannot be self-referential. In particular, the situation of self-referential assertions include their own linguistic expression (cf. Barwise/Etchemendy [1987: Chapter 8 and 9]).

situation the proposition is about. The reasoning used to analyze the Liar applies here too; we must pass from (S) to

(S') $s^{(2)}$: $s^{(1)}$ is E.

Doing so, the self-referential structure is apparently lost. This means nothing more than that the thing that refers and the thing that is referred to are not the same. In reality, $s^{(1)}$ and $s^{(2)}$ are not arbitrarily different things, but they are different representations of one and the same thing in reality which comprizes both in a union, building a whole.

Consider the following example:

(5) a: a is an English sentence.

In this case one has to distinguish three things. First, the interior "a" is the subject of a proposition which claims about a that it is an English sentence. Second, the exterior "a" designates this proposition. Third there is the claim that a is, in fact, an English sentence. Hence we are dealing with the question of whether the predicate expressed about a applies to a taken as the whole, that is, whether the conceptual content of the proposition applies to its linguistic representation. The concrete union of the conceptual content of this proposition with its actual linguistic representation is the same union spoken of in (S') above in greater generality.

This concrete union does not arise, if the conceptual content of the proposition does not apply to its linguistic representation, as for example in:

(6) a: a is a Chinese sentence.

However, both expressions (5) and (6) lead to a contradiction as before, if we ignore the different meanings of the interior "a" and the exterior "a". In particular, a symbol having distinct concrete significances is projected into the conceptual realm, leading to a contradiction (cf. section 6 and 7). The contradiction is more obvious in case of (6), since the exterior "a" quite clearly is not a Chinese sentence, but in principle the analysis of (5) and (6) is the same as before.

Note that (6) is merely self-contradictory in the sense that its propositional content does not apply to its linguistic expression: But it is not antinomical, since we cannot derive a logical equivalence of mutually negated propositions.

Consider now the self-referential statement (S) where self-reference itself is denied:

(7) s: s is not self-referential.

The expression (7) arises from (S) if we replace the property E by the negation of self-reference. The contradictory character of (7) is not surprising, because the structure of this assertion is actually self-referential while that is denied by the conceptual content of the proposition.

In (6) and (7) one has to observe *two* overlapping contrasts. First, there is the now familiar contrast between the interior and the exterior "a" or "s" respectively. They are distinguishable objects in the sense of section 2. Another contrast, however makes itself manifest in these examples. When the reader understands the proposition involved in (6) and (7) he notices a conflict between their conceptual content (meaning) and their very form (linguistic expression). The conflict in (6) arises out of the accident of what language is used. One can construct other such conflicts involving accidents of representation, such as:

a: a has fewer than four words.

The conflict appearing in (7) lies in the fact that the conceptual content of the proposition is at variance with the structure of the linguistic expression that carries it.

Note that this contrast between the conceptual content and the linguistic expression of a self-referential assertion also involves two distinguishable objects in the sense of section 2. The projection of *this* contrast into the conceptual realm leads to a contradiction in both cases (6) and (7). This is a common property of all self-referential assertions of the form (S).

According to Barwise/Etchemendy [1987] one may say that the following self-referential assertion,

(8) s: s is self-referential,

signifies its own situation. It is not only a proposition, but a *descriptive proposition* about a situation which it explicitly refers to, namely its own linguistic occurrence. However, this is also true for (6) and (7). The situation of (8) consists only of facts, in particular, of linguistic expressions. In this case, the propositional content, in fact, applies to the situation it is about. Hence, in this sense, (8) is true. As we shall see more clearly in the next section, (6) and (7) are false.

Note that self-referential *assertions* are not propositions in the usual sense but linguistic expressions of a certain structure which relate a propositional content with its own linguistic expression. We may call them *self-descriptive propositions*. In other words, they are propositions which are related uniquely to a specific situation, namely the situation which includes their linguistic expression. (For a more general definition of descriptive propositions in contrast to conceptual propositions, see section 9.)

The relation of self-referential assertions to their own linguistic expression cannot be made explicit in all its implications by any symbolic notation. For example, using the same symbols in (8) for both instances of *s* before and after the colon is misleading, but expresses the self-referential structure of (8) in a most natural way. Introducing different symbols for these *s*'s seems to destroy just this self-referential structure. The only sensible thing to do is to combine these two notions. This amounts to treating the two instances of *s* as different (more precisely, distinguishable) but also as equal in the sense that they are both manifestations of one and the same whole.

Ignoring these distinctions does no harm in dealing with self-referential assertions which are not antinomical.[7] We just loose some aspects which might be important to keep in mind if we want to understand the nature of self-reference in a deeper sense.

9. The structure of self-referential assertions

The considerations of the last section make us aware of the basic difference between the labeling part and the propositional part of a self-referential assertion. In particular, we have the linguistic expression and its intended conceptual content. Self-reference is only possible if the thing which refers (the referrer) and the thing which is referred to (the referent) are different. Otherwise there would be no reference at all, only monotonous identity. For instance, in (5), the conceptual content of the proposition and the linguistic expression are both representations of one and the same whole, namely the object under consideration. In effect, what has to be referred to each other are two representations of one and the same thing.

---

[7] There even are consistent formal theories which include self-referential structures which do not take explicitly into account all our distinctions. However, they must in one way or the other cope with the self-referential antinomies. See Barwise/Etchemendy [1987] for a particularly elegant treatment of self-referential assertions and their model theory and Aczel [1988] for the mathematics of self-referential (i.e. non-well-founded) sets. For a characterization of what makes a self-referential assertion antinomical, see Wandschneider [1993, §§ 3 and 4].

These two representations are in (5), and a fortiori in (S), in a special relationship. Namely, the actual sentence has a structure corresponding to the conceptual content of the proposition. It is, so to speak, an instance of the latter.

The structure of self-referential assertions is such that they are not propositions in the ordinary sense, where one has only to deal with conceptual entities; but they relate two representations of one thing, namely the actual instance and its corresponding conceptual structure. Propositions which connect concepts without refering to their instances will be called *conceptual propositions*. For example, the following proposition is conceptual: 12 is divisible by 3. As soon as we distinguish between two different representation of an object, we leave the purely logical realm. Because, logic deals only with *patterns* of representation, that is, *possible* instances and not, as it is the case here, with *actual* representations (or instances).

Hence, self-referential assertions are not part of pure logic: They are not conceptual propositions but so-called *descriptive propositions*. It is no wonder then that purely mathematical accounts of self-reference are wanting in one way or the other.

Descriptive propositions deal with the question of whether an observed real object is an instance of a given concept or not. In the first case one says that the concept applies to this object, or that this object is an instance or an exemplification of this concept. Self-referential assertions are *descriptive* propositions about their own linguistic expression, disguised as *conceptual* propositions about themselves. In order to make this more explicit, we consider the following new notation.

If we denote by "A" a concept and by "a" any of its instances, then we might use an arrowhead, $\triangleright$, for expressing the fact that a is an instance of A in the following way: $A \triangleright a$. This expression signifies the *union* between A and a which is neither identical with the concept A nor its instance a. Consider the following example for this structure: Take K as the concept of the sphere in 3-space. If k is a free flying soap-bubble, then we might write: $K \triangleright k$. Be prepared to differentiate clearly between K and k. The assertion $K \triangleright k$ means that an actual thing, called k, is an instance of the concept K. In this case, since K applies in reality to k, we might speak of the concrete union of K and k, constituting a whole.

Let us return to self-referential assertions. Their basic structure is expressed by

(S) s: s is E,

where E is any property which can reasonably be applied to a linguistic expression. Our earlier analysis shows that (S) is a self-descriptive proposition. In other words, (S) is a descriptive proposition such that its propositional content applies to its linguistic expression. Using our new arrowhead notation, this can be expressed by

$$E \triangleright \text{"}s \text{ is } E\text{"}.$$

Conversely, any descriptive proposition which has this structure can be expressed as a self-referential assertion.

The expression on the left side of $\triangleright$ always denotes a conceptual entity (in this case carrying propositional content); the expression on the right side of $\triangleright$ denotes a concrete object (in this case a linguistic expression) which is an instance of the conceptual content on the left side.

We can still go one step further in playing with the expression of self-reference. The following expression mentions self-reference and also *is* self-referential according to the definition in section 8. Its structure is an instance of its own propositional content:

(8) s: s is self-referential.

Now, "self-referential" can be replaced by "refers to s", giving

(9) s: s refers to s.

Once again the method of intrinsic analysis leads us to discriminate among the instances of s: Both occurrences of s on the right side of the colon are ingredients of the propositional part, they constitute its subject and predicate. The s preceding the colon designates this proposition itself. Hence we have

(9') s: $s^{(1)}$ refers to $s^{(2)}$.

It should be clear from the intrinsic analysis of (9) that its propositional part, namely

$$s \text{ refers to } s$$

is not a purely conceptual proposition. Because, in this case, the referrer and the referent would need to be conceptually different in order to express a relation and not a monotonous identity. In fact, this proposition is descriptive, namely self-descriptive, and it expresses the fact that a concept s refers to an instance of it, so using the arrowhead notation: $s \triangleright s$. In addition, according to (9), the very structure of s expresses self-reference. In other words, s also expresses the concrete union, or whole, of its constituent parts which consists of s taken as a conceptual entity (which is in this case the concept of self-referential assertions) and *s* as a concrete instance of its propositional content. Hence our symbolic notation produces

(10) $s: \ s \rhd s.$

Once again, the self-referential structure emerges instantly from (9) and (10), as long as we do not use different symbols for the three instances of *s*. But our previous analysis shows that the exploration of the fine structure of this self-referential assertion forces differentiations upon us which cannot be ignored unless we forgo, in effect, the very nature of self-reference. From this follows that we cannot stick to (10) but must introduce the differentiation used in (9'), giving

(10') $s: \ s^{(1)} \rhd s^{(2)}.$

In this expression, $s^{(1)}$ corresponds to the $s^{(1)}$ in (9') and represents the conceptual content of the propositional part of this self-referential structure. The concrete instance entering into this expression is represented by $s^{(2)}$. The fact that $s^{(2)}$ is a concrete instance of $s^{(1)}$ makes *s* a descriptive proposition, more precisely, a self-descriptive proposition.

The symbols $s^{(1)}$, $s^{(2)}$, *s* do not denote arbitrarily different things. We introduced these different symbols to aid our analysis. These three symbols refer to different aspects of one and the same whole, i.e., one and the same real object, namely the thing under consideration. In (10') we have a self-referential assertion with the property that its propositional content states exactly its own self-referential structure. Symbolically, *s* expresses the fact that it is self-referential, namely that $s^{(1)}$ applies to $s^{(2)}$.

It is now easy to represent in our new arrowhead notation a self-referential assertion which is contradictory:

(11) $s: \ s \rhd$ non-*s*.

(11) is equivalent to (7) and arises from (10) by negating the fact that *s*, taken as a self-referential assertion, is an instance of its conceptual content. In effect, the descriptive proposition in (11) denies explicitly its very self-referential structure. Therefore, the conceptual content of this proposition is incompatible with its actual structure expressed linguistically. In other words, (11) is structurally self-descriptive but the concrete instance does not match the conceptual content of the descriptive proposition. Hence, in this sense, (11), taken as a self-descriptive proposition, is false.

This analysis of contradictory self-referential structures is what Finsler [1925] and [1944, §2] might have had in mind, saying, that contradictory assertions of the form (11) are "solvable". In particular, he said that in every such expression the explicit conceptual content contradicts its (conceptually) implicit linguistic structure. However, from our point of view, there is no prima facie contradiction, but distinguishable objects in the sense of section 2 which produce a contradiction only if projected into the conceptual realm.[8]

10. Self-referential concepts

In an intrinsic analysis we are aware of our patterns of thought. We recognize different aspects of self-referential assertions as our mind shifts back and forth between the linguistic expression and its propositional content. This makes us aware that self-referential assertions are disguised as purely conceptual propositions: But we know, in fact, from observing our own thinking, that they are partly descriptive.

Prima facie, a self-referential assertion appears as an unstructured unity. Our analysis finds distinct components in this assertion. Our reflection discovers a complex unity comprising the components which our analysis has identified.

A severe challenge lies waiting to test the method of intrinsic analysis. There are, after all, the logical antinomies which seem to involve, according to section 1, only concepts and conceptual relations from mathematics and logic. The intrinsic analysis of self-referential assertions observed the mind at work, and identified a conceptual component (propositional content) as well as a perceptual component (linguistic expression). In all examples of antinomies discussed so far, the perceptual component is drawn on sense-perception. For instance, the left-hand labels in example (1) are visually associated with the subject of the proposition on the right side. Clearly, perception is at work here.

The logical antinomies, however, challenge us by lacking, prima facie, a perceptual component. They seem to be purely conceptual. But that does not mean, however, that an intrinsic analysis of logical antinomies is confined to purely deductive reasoning.

---

[8] There is a remarkable similarity between Finsler's treatment of the Liar-type antinomies and Buridan's ideas about self-reference (cf. Hughes [1982]). Buridan too does work within the realm of classical (absolute) logic and is convinced that every proposition is either true or false. Concerning the Liar, he comes to the conclusion, that it must be false. His argument rests upon the distinction between a sentence as a linguistic object and a sentence as a carrier of conceptual meaning. In fact, our analysis shows that this dinstinction is instrumental for the intrinsic diagnosis of self-referential assertions. Buridan was apparently aware of this fact.

The world of concepts can be taken as a landscape whose forms and relationships stand available for recognition (conception). It is true, we ourselves participate more in this recognition than in the perception of sensory objects. But, nevertheless, we single out concepts as objects of our attention, observe distinctions and relations, as we would with objects of the sensory world. However, concepts do not appear to us in their essential structure without voluntary thought-activity. They do not drop into our consciousness by themselves. Hence we recognize a part of this landscape only as we are *actually* thinking.

It is important to notice here, that we are now dealing exclusively with concepts and not with assertions, sentences, etc. That is, the subjects and predicates of all propositions we are going to analyze are themselves conceptual.

Let us begin with an analysis of self-referential concepts, called *predicative* concepts. We follow Grelling/Nelson [1908] in their account of an idea going back to Russell (cf. also Finsler [1927b]).

> Every concept can either be applied to itself or not. The former shall be called "predicative", the latter "impredicative". (Examples of predicative concepts are: conceivable, abstract, consistent, invariant, as well as all other concepts which denote essential properties for the concepts themselves; In addition, many negative concepts are predicative, as for example, non-human, etc. The following are impredicative concepts: virtuous, green, and most of the everyday concepts.)[9]

Let us look at one of these predicative concepts more closely. The concept of abstractness means nothing else than the essential structure of all abstract objects. Let us call this structure "S_A". Hence S_A is the essential structure of all concepts which have the property of being abstract. Let us now single out a specific abstract concept as an object of our attention; the concept of the Riemann integral from calculus will do nicely. The Riemann integral, let us call it $S^{(i)}$, has an abstract structure, is an instance of S_A, therefore in the light of our previous considerations, we have

(12) $$S_A \rhd S^{(i)}.$$

On the left hand side is a conceptual category as before, namely the essential structure of abstractness, S_A. On the right hand side, however, there is no longer an object of the sensory world, as the

---

[9] Grelling/Nelson [1908: 60f.]. Translation by R. Ziegler/D. Booth

linguistic expression in (10'), but a conceptual entity. This conceptual entity, $S^{(i)}$, stands as a specific instance of the conceptual category S_A in the same manner as linguistic expressions were specific instances in our earlier analysis of self-referential assertions. In other words, (12) is a descriptive proposition about the specific concept $S^{(i)}$.

To obtain a self-referential concept of the type treated by Grelling/Nelson, we must now take S_A to be an instance of itself. Abstractness is itself abstract, or in symbols:

$$(13) \qquad\qquad S_A \triangleright S_A .$$

Now, in this notation, the referrer and the referent seem to be one and the same. Thinking about this we realize that there is an actual distinction: The referrer and the referent arise differently. Were they undifferentiated, there would be no reference at all, in particular no self-reference. To interpret (13) consistently, the left side must be taken as a conceptual category and the right side as a specific instance of it.

Taking S_A as an instance of itself, is tantamount to assigning S_A a different quality beyond its conceptual structure or meaning. This structure appears as a part of the conceptual landscape, having its own ontological substance without which it has no existence. Any quality which we state *about* concepts (as, e. g., the abstractness of the Riemann integral) is not an essential part of their structural content, and hence not relevant for purely logical or mathematical considerations: But it is essential to their form of existence or appearance. The analysis of self-referential concepts shows that, in general, on cannot talk *about* concepts while denying them any kind of ontological quality: There would be nothing left to talk about.

Now let us turn to the concept of predicativity itself as described by Grelling/Nelson [1908]. In general, a concept C is called *predicative* or *self-referential*, if

$$(14) \qquad\qquad C \triangleright C .$$

The distinction between the referrer and the referent is made explicit by our method of separating parameters:

$$(14') \qquad\qquad C^{(1)} \triangleright C^{(2)}.$$

The concept of predicativity (or self-reference) contains an element not found in the concept of abstractness. The actual relationship between the conceptual category and one of its instances is the essential structural part of the concept of predicativity; it is not essential (in fact accidental) to the concept of abstractness. When Grelling/Nelson introduced the concept "predicative", they lead us to single out this relation. In fact, this relation is exactly what constitutes the concept of predicativity. Our intrinsic analysis therefore requires that we introduce a third parameter which expresses the fact that the relation between the constituents of the concept predicative, namely the referrer and the referent, is but a different representation of these constituents themselves:

$$(15) \qquad\qquad\qquad C\colon C^{(1)} \vartriangleright C^{(2)}.$$

Without this third parameter C, we would only have a predicative (self-referential) concept and not the concept of predicativity itself.

Our analysis of predicative, or self-referential, concepts produces the same patterns here in (14) and (15) as were obtained in the analysis of self-referential (linguistic) assertions. It is truly remarkable that these patterns are alike; for in our analysis of the self-descriptive statement (10') we were moved by noticing a *perceptual* aspect in the statement. No actual perception arises out of the concept of predicativity. As we ponder predicative concepts we need to isolate aspects of them for our attention just as we need to distinguish the perceptual and conceptual components in self-referential assertions. From this observation we obtain two of our three parameters. These stand for *distinguishable* objects in the sense of section 2, where one has a property the other lacks. This distinction plays an important role in the following analysis of logical antinomies.

11. Diagnosis of logical antinomies

Having analyzed self-referential concepts in general, let us now turn to the analysis of a logical antinomy. The most direct of the logical antinomies is that which Grelling/Nelson [1908, 60f.] report as being from Russell. We shall continue the quotation begun in the previous section.

> The concept *impredicative* is itself either predicative or impredicative. Assume that it is predicative, it follows from its definition that it is impredicative. Assuming that it is impredicative, it does not apply to it itself, hence it would not be impredicative. Both assumptions lead to a contradiction.

The very concept of impredicativity is the focus of this antinomy. Any concept, C, does apply to itself, i. e. $C \vartriangleright C$, or does not appply to itself, $C \vartriangleright$ non-C. In the former case it is said to be predicative or self-referential; in the latter impredicative or non-self-referential. The question is, whether the concept impredicative, let us call it C_{im}, is predicative or not. The concept impredicative comprizes all concepts C which have the property of being impredicative, i. e. all concepts C with $C \triangleright$ non-C.

The derivation of the antinomy using the present notation proceeds in the following way. In case C_{im} is actually predicative, we write $C_{im} \triangleright C_{im}$. But because of the meaning of C_{im}, we would then have: $C_{im} \triangleright$ non-C_{im}. On the other hand, in case C_{im} actually is impredicative, one has $C_{im} \triangleright$ non-C_{im}, so, taking in account the meaning of C_{im}, we have: $C_{im} \triangleright C_{im}$.

In deriving this contradiction, we begun with the assumption that $C_{im} \triangleright C_{im}$. This assumption is not merely formal, but has actual consequences. The very meaning of C_{im} is that of concepts being not self-referential. So the antinomical reasoning leads us to $C_{im} \triangleright$ non-C_{im}. Were we to begin with the opposite assumption, $C_{im} \triangleright$ non-C_{im}, a parallel chain of thought produces a contradiction too.

An intrinsic analysis of this argument leads us to separate parameters, so that we can go beyond the superficial pattern of the argument. The argument heads of with the assumption that C_{im} is actually impredicative, i. e. $C_{im} \triangleright C_{im}$. Right here, however, we have to distinguish between the conceptual content of C_{im} on the left and the concept of C_{im} as an object, a conceptual entity on the right, giving

$$(16) \qquad\qquad C_{im}^{(1)} \triangleright C_{im}^{(2)}.$$

This expression says that $C_{im}^{(2)}$ is an *instance* of $C_{im}^{(1)}$, that is, the concept impredicative, taken as an *object* (conceptual entity), has to satisfy its own conceptual content. Hence we have

$$(17) \qquad\qquad C_{im}^{(2)} \triangleright \text{non-}C_{im}^{(2)}.$$

The things denoted by $C_{im}^{(2)}$ in (16) and (17) are distinguishable objects in the sense of section 2. In (16), $C_{im}^{(2)}$ is a specific conceptual object, which is an instance of the conceptual content $C_{im}^{(1)}$, whereas in (17), $C_{im}^{(2)}$ itself is a conceptual content of which non-$C_{im}^{(2)}$ is an instance.

Conjoining (16) and (17) gives a contradiction if and only if $C_{im}^{(1)}$ and $C_{im}^{(2)}$ are identified. In identifying these parameters, however, we loose the very distinction which allowed us to make the transition from (16) to (17) in our chain of reasoning.

A similar argument applies to the chain of reasoning starting with the opposite assumption, $C_{im} \triangleright$ non-C_{im}. We conclude that this antinomy is equivalent to neglecting the distinction between the

content of a concept and the concept as a substantial object
(conceptual entity or instance) in itself.

One might think that this distinction amounts to a typed
response. However, type theory does not explain or actually solve an
antinomy, it simply avoids them. In contrast, our approach goes
back to the roots of logical antinomies and thus reveals the reason
why they arise in the first place.

This paradox concerning the concept impredicative seems to be
an antinomy only to the thinker who neglects the principle of
identity by confusing the content of a concept with the concept as a
substantial object (not from the sense-perceptual reality, of course),
i.e. the concept as an *instance* of another (here, the same) concept.
In other words, a pragmatic contradiction (in our earlier
terminology, distinction) is projected onto the field of conceptual
relations, producing what appears there to be an analytical
contradiction.

In his discussion of the antinomy concerning the concept
"impredicative", Finsler [1927b] stresses the fact that we cannot
expect circular or self-referential definitions to be satisfiable in any
case. There are exceptions, and the definition of "impredicative" has
the property that it cannot be applied to itself without producing a
contradiction. Therefore, applied to itself, the concept "impredicative"
has no meaning.

Russell's set theoretic paradox arises from the antinomy of the
concept "impredicative" if we switch to the extensional point of view.
Sets which do contain themselves as elements ("predicative sets") are
not well-founded. Sets which do not contain themselves
("impredicative sets") are well-founded; they are also called "normal".
The set of *all* normal sets is Russell's set R. If R is normal, it does
not contain itself as an element hence R is not normal. Conversely, if
R is not normal, it does contain itself, hence it is normal. This is the
antinomy.

Finsler [1926b], [1927b] draws from this situation the same
conclusion as above: Self-referential definitions need not be
satisfiable.

Conclusion

All references involve the distinction between a refering part
(referrer) and an object which is referred to (referent): This
distinction is easily forgotten where self-reference is concerned. But
in reality it is still there. Were we to neglect this distinction, there
would be no actual reference at all, merely an identity. In deriving a

contradiction, using one of the self-referential patterns discussed earlier in this paper one actually recognizes first of all the distinction between the referrer and the referent. The given distinct objects have to be recognized as real. But, in the next stage of the argument, the distinction is dropped, an identity is assumed, and the illusion of a contradiction is produced. Hence, the antinomical argument produces an illusory contradiction from a real distinction, based upon a violation of the principle of identity.

Within the antinomical patterns, there lurks a contradiction waiting to catch us if we neglect the deeper aspects of self-reference. The need for intrinsic analysis of the deeper aspects of reference exist in any instance of self-reference, quite apart from whether an antinomy arises or not. The antinomies merely make us aware of something that is present in all self-reference. Hence the heart of the problem is the structure of self-reference.

When there is self-reference, the referrer is a concept and the referent is an observed instance, so observation is necessary for self-reference. For self-referential assertions, like the truth-teller or the Liar, which involve actual linguistic expressions, the observational part of the reference is directly related to sensory perception. For self-referential concepts, however, the principle is the same, but the observation is something deeper and more elusive, an observation *within* the conceptual realm. This is a necessary consequence of our accepting the existence of self-referential concepts. This observation, a primary experience by its very nature, reveals a substantial component, namely some kind of ontological basis of conceptual entities.

Thus it turns out that the question of whether concepts have an existence independent from our mind is not a purely theoretical one, subject only to individual beliefs. This would mean also that this question cannot be decided by deductive reasoning from some "well-known" principles. On the contrary, as we have shown, what is needed is the observation and analysis of our own thinking process. The basic observational facts and experiences must be drawn from carefully designed mental experiments. In this paper, one exceptionally rich example was presented and analyzed, namely the concept of self-reference and some of its manifestations.

Someone who attempts to deny this observational (but *not* sense-perceptual) component of concepts as an objection to these views, must exclude all self-referential concepts from his considerations. In fact, this was done by Russell [1908], who explicitly eliminated all self-referential structures. However, this cannot be done consistently, as was shown conclusively by Várdy [1979]. Suppose, self-reference to be forbidden: No assertion or concept applies to itself. Obviously, this assertion has a self-referential structure and denies in its propositional content all self-reference, in particular its own. This leads to an antinomy.

An intrinsic analysis of semantic and logical antinomies shows that they, in fact, rest upon the same structure. In deriving the contradiction, one has to violate the principle of identity, while identifying the referrer and the referent. Therefore, from this point of view, semantic and logical antinomies are similar. However, they differ in the realm from which the referent is drawn. The referents in the semantic antinomies are drawn from perception. The referents of the logical antinomies, on the other hand, are drawn from the realm of pure concepts, hence from conception. It is obvious, in the case of semantic antinomies, that the specific perceptual reference transcends logical forms. Even the logical antinomies contain specific references which are not part of formal logic, though they remain accessible to human reason. Pure logic, after all, concerns itself with *possible* objects: The distinctions that underlie the antinomies ultimately refer to *actual* objects. In other words, our intrinsic analysis of logical antinomies concerns itself with the transition from conceptual propositions to descriptive propositions within the realm of pure concepts. Self-reference is the key which opens our minds to the experience of the ontological status of concepts.

Acknowledgements

This paper is the result of many discussion I had with Werner A. Moser and owes much to his clear, crisp and yet very experiential approach to systematic philosophy. For a more elaborate discussion of some of the topics developed here, see the book Ziegler [1995]. Additional helpful comments came from Bernd Gerold, Thomas Kesselring, and Thomas Meyer. Discussions with David Booth while translating this paper together resulted in many improvements, both conceptually and linguistically.

Inaugural lecture, University of Köln, 1923. First published as: "Gibt es Widersprüche in der Mathematik?" *Jahresberichte der Deutschen Mathematiker-Vereinigung* **34** (1925), 143–155.

Are There Contradictions in Mathematics?

Can contradictions exist in mathematics? That is, *insoluble* contradictions? In this, the most exact of the sciences, is not every statement *either true or false*, quite independent of all personal opinions, points of view, or other influences? Is it possible to prove a proposition and also at the same time its *negation*?

Some think that these questions are superfluous. Mathematics has had the reputation since ancient times of being *absolutely true* and *indubitably correct*. Of course, false deductions, errors, or miscalculations can happen but these can nevertheless be avoided by paying sufficient attention. But to arrive at contradictions *without* having made fallacious deductions: Surely this must be excluded.

And now, this inner consistency of mathematics *has* been thrown into doubt!

Already in antiquity Zeno believed that genuine contradictions existed in the theory of motion and that this whole science had fallen into absurdity because of them. The arguments he gave were intuitive versions of purely mathematical questions and posed very serious problems for the contemporary theory. Today, however, we possess rigorous concepts of convergence and continuity, so that these things can be considered mathematically clarified.

Later, especially when mathematics received fresh impetus after the discovery of the calculus, people troubled themselves little about its rigourous foundations. To obtain results was what carried weight; whether or not the way in which these results were obtained was completely free from error, was a matter of lesser importance.

So it came about that real errors in reasoning were unconsciously introduced; but a feeling of certainty about the results gave mathematicians confidence.

When the fallacious deductions that had been made became apparent, the lack of a sound foundation for the whole of mathematics, one that would afford the greatest possible certainty against any errors whatever, made itself felt.

Yet, just when it was believed that this goal had been reached and when, furthermore, it had come to light that one can also calculate with infinite numbers, i.e. with sets of infinitely many things, in an unambiguous way, the investigations ran up against some of the strange contradictions which threaten the whole foundation. Right up to this day no generally accepted explanation has been found.

We must therefore occupy ourselves with these *antinomies* if we want to know whether, or to what extent, mathematics can be considered free of contradictions.

I should like to illustrate the nature of these contradictions with two examples, not very difficult ones, both of which are due to Russell. The simpler is more purely logical: The other stems directly from the foundations of set theory and touches upon the foundations of mathematics.

The first is connected with the following question: *Which is the smallest natural number that cannot be defined with fewer than 100 syllables in the English language*? And above all *does there exist* such a number?

It can be shown that such a number must exist. For, among the natural numbers there are certainly finitely many numbers each one of which can be defined with fewer than 100 syllables. There are only finitely many syllables, so there can only be finitely many combinations of 100 syllables or less. Of these combinations only a small portion represent meaningful definitions of numbers. There are still other numbers which cannot be represented in this way, and among these there must be a smallest. This, then, is the number that is sought.

On the other hand, one can reason that there is no such number. For, should it exist, one could define it with fewer than 100 syllables by means of the phrase: "The smallest natural number that cannot be defined with fewer than 100 syllables in the English language." A contradiction results.

The second example is that of *the set of all sets that do not contain themselves as elements*.

If any objects whatever are given, then one can collect them together into a *set* and call these objects the *elements* of the set. One can thus speak of the set of all men; each man is contained in this set. He is an element of the set. Similarly, one can form the set of all numbers, the set of all circles in a plane, and so on.

These are examples of sets which do not contain themselves. The set of all men is not itself a man, it is therefore not identical to one of its own elements. Similarly the set of all numbers is not itself a number, so it is not contained in itself. Whether there also exist sets which do contain themselves may be left as an open question for the time being, and is in any case of no consequence for the present line of thought. It is enough that there do exist sets which do not contain themselves. Now, combine all these sets into a new set. They form "the set of all sets that do not contain themselves".

But now, how do things stand with this set? Does it contain itself or not?

Suppose that it were to contain itself. Then it would be a set which contains itself as an element. It ought, however, to contain *only* such sets as do not contain themselves. Therefore it cannot contain itself.

On the other hand if it does not contain itself, then it would have to be one of those sets that do not contain themselves. It ought, however, to contain *all* the sets of this kind and therefore it would have to contain itself.

Thus, if it contains itself, then it is not allowed to contain itself: And if it does not contain itself, then it must, all the same, contain itself. This is a contradiction.

This contradiction is not without significance for *set theory*. Set theory is a comparatively new branch of mathematics; it was founded by Georg Cantor a few decades ago and has in the meantime attained fundamental significance for many parts of mathematics. An attempt was made to build up the whole of mathematics from set theory but this has run aground on precisely these antinomies. Of course, it is possible to form particular sets which appear to be quite free from objection and one can construct sets of ever greater size but the question is: *How far* can one proceed in this way? In the end one certainly does run into contradictions.

Further, if it is not possible to collect together all the sets with some definite property, is one then allowed to speak of all the numbers which possess some definite property? Are contradictions possible in such a case too? But then, reasoning of this kind is in continual use throughout mathematics.

Many attempts have already been undertaken in order to *solve* or *avoid* these difficulties; here I can briefly characterize only the most important of these.

Doubtless many view the first of the above mentioned examples as being not particularly serious and content themselves with any one of the plausible explanations. The most widely accepted solution is, of course, that it is only a question of the *inexactitude of language*. Language is not logically exact; the number of syllables in a definition is not a mathematically legitimate quantity.

To this view it has to be objected, however, that it must be shown where the inexactitude is to be found and whether it can then be corrected so that such things can no longer occur. For, if ordinary language were really to lead into insoluble contradictions, then it would no longer serve for general use.

The attempt at a solution by Henri Poincaré is not very different from this. He says that a classification of something according to the number of syllables of the sentences used to define it is *not well defined*. For, during our inspection of the sentences some of them, namely those which depend upon the classification itself, may change their meaning. Therefore the classification is unstable and can never be fully completed.

Yet how is it if one restricts oneself to *entirely unambiguous* sentences? If I separate off only those numbers which can be defined in a completely unambiguous and unvarying way with fewer than 100 syllables in the English language, then this classification is certainly well-defined. But then again there must exist among the remaining numbers a smallest one which is then determined in an entirely unambiguous and unvarying way. The contradiction has not been removed.

Russell, who seeks to get rid of the various contradictions with the help of his *theory of types*, goes about the problem in another way. A similar theory is also due to Julius König.

What Russell says in relation to the second example is roughly the following: If one collects together sets of a specific type then this gives rise to a set of a new type. It is not permissible then to collect together sets of different types.

With that the contradictions will be avoided; but also only avoided and not solved, for, it is really by no means evident why one should not also be able to collect together sets of different types. But as soon as one has the freedom to collect together arbitrary types, the contradictions arise anew. Besides, the concept of the type, even should it turn out to be useful, is difficult to define; according to Russell's own words, a large part of the theory is still to this day chaotic, confused and dark.

Recently two mathematicians, namely Brouwer and Weyl, have even gone so far as to reject the *law of the excluded middle*. They say, for example, that two numbers are not necessarily either equal or different. There is still a third possibility. It could be that the two numbers are neither equal nor unequal but in some way *indistinguishable*.

In themselves such assumptions may lead to quite interesting investigations, but *exact* science can certainly not be founded upon them, quite apart from the great complication which would arise; also many of the most secure results would have to be abandoned.

Now in order to rescue at least that part of set theory which is important for applications, Zermelo has set up an *axiom* system which exclusively determines the formation of sets. As far as it goes, however, this can only be viewed as an emergency measure. Zermelo shows that he avoids the well known paradoxes but does not show that contradictions really are impossible in his restricted domain.

Finally Hilbert, the main advocate of axiomatic method, seeks to *prove* the consistency of the separate parts of mathematics. He shows in his "Grundlagen der Geometrie" [1913] that elementary geometry can be completely reduced to the arithmetic of the real numbers. Every contradiction that is to be found in geometry must appear also in arithmetic, and thus if the former is to be consistent then so must be the latter.

But how is one to demonstrate the consistency of arithmetic? One must eventually return, via set theory, to *pure logic*. Yet it is just here that the difficulties arise, for here one runs up against the antinomies.

Hilbert also sought, therefore, to axiomatize and formalize logic. He developed logic and arithmetic together, step by step, in order to prove the consistency of each step. The original ground upon which he builds everything is the *recognition of symbols*.

This of all things seems to me to be dubious. There is something of a subjective nature surrounding the process of recognition. Does it make sense to consider the recognition and combining of mathematical symbols to be more reliable than pure logic? So that one can base logic itself upon it?

Quite apart from this, however, and with all due recognition of the endeavours undertaken and the results achieved by Hilbert in this field, one must surely say that, in order to "restore the old reputation of incontestable truth to mathematics", such proofs alone cannot suffice. For if there exists a contradiction which really is inexplicable in even one part of mathematics then strictly speaking all proofs are threatened. One could never be certain whether or not an inexplicable contradiction might arise within what seemed to be the most secure arguments. It serves *no purpose whatsoever* to say that the contradictions occur in the boundary regions of mathematics. For where in set theory does the boundary of the boundary regions lie? Do not the investigations of Hilbert too take place in such a boundary region?

No, the *only* path which can lead to the goal is the one which really clarifies the antinomies and really solves them.

But can we travel such a path?

After many fruitless attempts the view has been expressed that a solution by means of an explicit presentation of the errors in reasoning is not possible. One will just have to put up with these things or at least await a transformation of logic.

But can one put up with contradictions in mathematics? As has already been mentioned all proofs would then loose credibility; even more it can be shown that any one contradiction entails all other contradictions. If one allows only a single contradiction then one can prove everything true and everything false in the same way that one argues from a false hypothesis. With that, however, science becomes impossible.

And what about a transformation of logic? Is it possible to transform logic? This could only be possible for a written or formalized logic. Such a logical system could be incorrect or too restricted: But this is not true of pure logic, *the* logic to which one *has* to submit as a rational being. Should this logic be changed, or were it to lead to contradictions, then again all science would be impossible.

Thus if one does not want to abandon all science, one has to conclude: that the contradictions are soluble.   This path must be open.

I will now attempt to proceed along this path. I begin with the simplest starting point.

The first example mentioned above is still somewhat complicated; in particular, it is not easy to see specifically which number is involved. I will therefore strip away everything that is superfluous retaining only the essentials.

First, I write on this blackboard the numbers 1, 2, 3 and then the sentence: "*The smallest number which is not specified on this blackboard*" (see figure 1).

1,   2,   3,

The smallest natural number which is not specified on this blackboard.

Figure 1

Does there exist such a number? One can now repeat the earlier argument: Only finitely many numbers are specified on the blackboard. Among the remaining numbers there must exist a smallest. On the other hand, if it did exist then it would certainly be specified on the blackboard by means of this very sentence.

Now, however, the situation is so simple that we certainly must find the solution. In fact if one proceeds purely logically then one comes to a completely *unambiguous* result. That, after all, is the goal: Every question must get an unambiguous answer.

It is good not to take these simple things too lightly. They are necessary for an understanding of the more difficult set-theoretic questions, in which reasoning of the same kind is found.

How is it now? I shall reason in a way that is common in mathematics. The question is: Which numbers are specified on this blackboard?

The numbers 1, 2, and 3 are there. Is the number 4 also specified on the blackboard? Suppose that it were, then it could only be given by means of the remaining sentence. This sentence requires, however, that the number under consideration should not be specified on the blackboard. This requirement would then contradict the supposition, which would therefore be false. It follows that the number 4 is not specified on the blackboard.

On the other hand, there are numbers which are not given on the blackboard; among these there must be a smallest. We have just seen that the numbers 1, 2, and 3 are given on the blackboard and that the number 4 is not given. Therefore the number 4 is actually "the smallest natural number which is not specified on this blackboard".

Is this really a contradiction? Does not this sentence actually specify the number four?

Answer: No, it does not. For, this sentence *as it stands* not only requires explicitly that the number in question should not be specified on the blackboard, but it also requires at the same time, implicitly, through the fact that it stands itself on the blackboard, that the number it specifies surely is specified on the blackboard. The number 4 does not satisfy these two self-contradictory requirements.

The essential *error* in reasoning which occurs in the usual treatment lies in the fact that the implicit requirement, in this case the requirement that the number certainly must stand on the blackboard, is only subsequently taken into consideration and not, as it has to be, taken from the outset. There is a difference between the sentence as it stands on the blackboard and the same sentence when it is spoken. The spoken sentence does not contain the implicit claim, whereas the sentence on the blackboard does contain it.

A good natured reader could perhaps say: The number 4 surely is intended by this sentence. Logic, however, is not so good natured. Taken logically this sentence is actually identical to the following: If the number 4 is not specified, then the number 4 shall be specified: If the number 4 is specified, then the number 5 shall be specified.

In this form it is quite clear that no number at all is specified by means of this sentence, neither the number 4 nor any number.

Similar paradoxes resolve themselves in the same way. The set-theoretic ones are somewhat more difficult.

It becomes apparent that the set of all sets which do not contain themselves as elements cannot really exist at all.

One could now argue as before: The contradiction is hidden in the definition. The definition requires something impossible. This set cannot be consistently defined; and something that cannot be consistently defined need not exist.

This explanation is indeed correct but not satisfactory. There still remains the question, and this is the heart of the set-theoretic antinomies: *How does it come about, or how is it possible that there do exist things which cannot be collected together into a set?*

We must ask ourselves: What is a set? According to Cantor it is "the collection of well-defined objects into a whole". That is to say, the collection itself. It really does seem as though one can always collect arbitrary things together to form a whole. What should there be to hinder one from doing this? Here, however, we have to be careful.

Yes, if one is dealing with concrete, material things, or those things that have nothing to do with the formation of sets, then there are indeed no reasons why one should not be able to collect them.

It is quite otherwise, however, when the things or their membership relations can enter into the formation of the set itself. These relationships could then be of such a kind that it is impossible for them to be fullfilled.

More precisely: If one forms sets of sets, or permits things dependent upon sets as elements, then the definition of Cantor is a *circular definition* and as such need not always be satisfiable.

As this is important, I should like to illustrate it by means of a simple example.

One can define a number x by means of an algebraic expression, for instance by the expression $a - b$:

$$x = a - b.$$

If a and b are fixed numbers which do not depend upon x then a circle does not arise. The definition is always satisfiable: There always exists a number x which satisfies it.

Now, the algebraic expression by means of which x is defined could, however, depend upon x itself. It is then a circular definition and can still, under certain circumstances, be uniquely satisfiable. If, for instance,

$$x = a - x,$$

then this condition is satisfied uniquely by the number $x = a/2$.

A circular definition does not, however, need to be satisfiable; in

$$x = a + x$$

when a is different from zero, there is no number satisfying it.

One cannot simply prohibit the circle; such equations occur often enough in applied mathematics and one must know how to deal with them.

It is exactly the same with sets. Cantor's general definition is a circular definition, so it is not to be wondered that there are instances in which it breaks down.

The definition on the blackboard (see above, figure 1) is also circular. The number to be defined is made dependent on the definition of this same number, and this dependence is of such a kind that it cannot be satisfied.

The antinomies are of a circular nature and the contradictions have to do with this fact; but this knowledge is of course not new. Yet, having recognized this, people then sought to avoid all circles by means of types or hierarchies, and were unable to attain the goal by these means. The proper solution is one which does not begin by attempting to avoid all circles. One must be far more clear about this point: A circle may very well be uniquely satisfiable, but this need not always happen.

There is still an objection which needs to be discussed. It is said: If one thinks of all really existing sets which do not contain themselves as being given, then one can certainly conceive that all these can now be again collected together into a whole, into a set.

We must reply: This is not true! One cannot conceive of this. In doing so one conceives of something different and beside the point. For, if one could conceive of this, then one would also have to be able to say whether or not this very collection is to be included. Does one conceive of this set as one which contains itself or as one which does not? One cannot say. No, something which is logically inconsistent can never be conceived; one can just as little conceive of this set as, for instance, of the largest natural number, which (as we know) does not exist either.

For a full treatment of the questions dealt with it is not only necessary to reveal the errors in deduction but it is also necessary to be able to show how they can be avoided; how, in particular, it is possible to achieve a *useful* and *consistent set theory*.

On this matter, however, I will have to confine myself here to indicating only a few essential points (see Finsler [1926b] for more details).

In set theory it is desirable to have the *system of all sets* as a *unique* and *fixed* system, as is the case for the system of all natural numbers in number theory.

This goal, which is not reached by the axiom system of Zermelo, can be attained if one introduces *only* one particular restriction which in fact allows sufficient scope for all applications, even though it may seem narrow.

First of all, in order to obtain a really fixed system, one must exclude all non-mathematical objects. At this stage the set of all men, for instance, cannot occur. The domain of arbitrary mathematical objects is also not sharply enough circumscribed, therefore it is still necessary to exclude even these as members of sets. All that remains are the *pure sets*, i.e., those sets whose elements are themselves again pure sets.

Such pure sets do exist, for instance the empty set {}, the set which possesses no elements; it plays an important role in set theory. It contains no element at all, hence nothing that is not a pure set. Moreover one can form the set which contains as its sole element the empty set {{}}; this is likewise a pure set. Further that set which contains the set just defined, {{{}}}, also that which contains both of the first two sets {{}, {{}}}, etc. Arbitrarily many infinite sets can also be produced this way.

For applications one can then represent other sets by such pure sets or, in as far as it is just a question of mathematical objects, define them directly in terms of pure sets. The sequence {}, {{}}, {{{}}}, ... , for instance, can be considered as being the representation of the sequence of the natural numbers.

The *totality* of *all* these pure sets, which, however, I have not defined completely, now forms a fixed system. (The complete "axiomatic" as opposed to "intuitive" definition is to be found in Finsler [1926b].)

But now the very same objections that are directed against the *set of all sets* and against the *system of all things* have also been raised against the totality of all pure sets. All these concepts are frequently considered as being equally contradictory.

These objections are not easy to refute; essentially they amount to maintaining that one can still enlarge the system of all sets. But this system, just because it does contain *all* sets, must be the largest.

A more detailed examination shows, however, that here again lie hidden fallacies of the sort discussed above. It is actually not possible to enlarge the system of all sets, and it is as impossible in this case as it is to specify on the blackboard a number which shall not be specified on the blackboard, or, just to mention briefly another paradox, as impossible as it is to define, with finitely many words, a number which cannot be defined with finitely many words, even though it can be proved with a fixed underlying dictionary that such numbers really do exist.

Within the system of all sets, as we have seen, it can happen that certain sets cannot be collected again into a set. In attempting certain collections *difficulties* appear to arise. One must first have a proof of the existence of each separate set.

In reality, however, these difficulties are confined to only a part of set theory, and indeed to only that part which, in any case, is not important for applications. That is to say, one can distinguish between *circle-free* sets, which can be defined free from circularity, and *circular* sets, for which this is not possible.

To the *circular* sets belong, in particular, every set which contains itself, but also, for instance, the set of all circle-free sets is circular. In the domain of these circular sets the usual operations are not valid, nor are the axioms of Zermelo. The paradoxes lie in this realm.

On the other hand the *circle-free* sets are those upon which the rest of mathematics can be built. If one only admits that circle-free constructions are always feasible, and this must surely be viewed as self-evident, then one can secure clarity concerning not only the questions of the consistency of arithmetic and analysis but also concerning such topics as the transfinite ordinal numbers and similar, hitherto controversial subjects.

In any case, the result that all contradictions are really only apparent does, I believe, stand firm; mathematics as such is free from contradictions. There is still a science in which nothing is valid except the pure truth.

First published as: "Formale Beweise und die Entscheidbarkeit", *Mathematische Zeitschrift* 25 (1926), 676–682.

Formal Proofs and Decidibility

I. Statement of the Problem

§1. In order to demonstrate the consistency of certain axiom systems, Hilbert makes use of a theory of mathematical proof in which the proof must be thought of as rigorously formalized in concrete symbols (see Hilbert [1922], [1923], [1926], Bernays [1922], Ackermann [1924]). "A proof is an array which must be graphically represented in its entirety" (Hilbert [1923, 152]). He adds: "A formula shall be said to be provable if it is either an axiom, or arises by substitution into an axiom, or is the concluding formula of a proof" (ibid., 152–153). The aim, then, is to show that, in a given axiom system, a contradictory formula (formalized in the same way) can with certainty never be proven. Axiom systems for which this can be demonstrated are said to be "consistent" (ibid., 157 and [1926, 179]). In the following where the formalization is quite general, such systems will be called *formally consistent*.

§2. The question of whether or not formal consistency implies a real freedom from inner contradiction in the axiom system is connected with the decision problem.

Were every mathematical assertion, whose truth or falsity is a logical consequence of some axioms, to be decidable by means of purely formal proof, then the problem of whether the system is consistent or not could be completely determined on the basis of formal consistency alone.

This no longer holds, however, if there exists a proposition whose validity is indeed logically established in a theory but which, purely formally, can neither be proved nor disproved. One could then add this proposition, or even its negation, to the system as a new axiom and thus obtain two formally consistent axiom systems at least one of which is internally contradictory.

§3. Now this possibility does in fact arise in connection with the axiom system for the real numbers, provided that there are at most countably many symbols, which is indeed certainly the case for concrete symbolic proofs.

There can then exist only countably many formal proofs, since each one must consist of a finite collection of symbols (Ackermann [1924, 9]). In the theory of the real numbers, however, there are more than countably many propositions. Consider the proposition: "α is a transcendental number". For each fixed value of α this is a definite, true or false proposition. Not all of these uncountably many propositions can be formally decidable. Otherwise there would have to exist uncountably many distinct proofs, even if some were variations of a general method. From this, however, it follows that there do exist propositions which are not formally decidable.

§4. It is still conceivable, however, that all propositions which can be represented formally are formally decidable. In this case the newly added axiom would not be formally representable. (One cannot formally represent quite arbitrary things simply by introducing a new symbol for them: One has to proceed in a fixed language and must carry out everything *formally*, the definition of new symbols included, otherwise the representation would not be *purely* formal – cf. end of §7.)

One could even suppose that such a non-formal contradiction might never be noticed; one could continually act as though it was not present. Possibly, no damage would be done.

It will now be shown by means of an example that it is actually possible to construct propositions which, in a general way, are not formally decidable. They are formally consistent but, all the same, a contradiction can be found by an informal argument. Hence it follows that the proof of the formal consistency of a system gives no guarantee against an actual contradiction.

These things are closely connected with the "paradox of finite definability", so a precise statement and explanation of this paradox will be given first of all (cf. Finsler [1925]).

II. The Paradox of Finite Decidability

§5. Let us assume that a fixed language L, consisting of finitely many symbols, is given. (One could also accept a countable infinity of symbols without causing any essential alteration.) In particular, let this language contain all the symbols necessary for mathematical purposes, or even, all the symbols ever used in writing or printing (perhaps even in the future). A fixed, "alphabetical" ordering of these symbols will also be taken as given.

Further, one forms a fixed dictionary, D, including a "grammar", which unambiguously gives meanings to certain "words" that are finite combinations of the symbols. Words with finitely many different meanings, as happens in common language, shall be unambiguously distinguished by means of indices. Certain symbols or combinations of symbols will be taken as "variables", which take their meaning from the grammatical structure and the context in which they appear. In particular, all the words of the foregoing discussion, with their understood meanings are in D, along with all words which have ever been printed or written until now (or even in the future). The "definitions" in the dictionary need not be expressed in L, they could be thought of as being given purely ideally.

Every combination of symbols having no unique meaning in D shall be held to be meaningless.

An object will be called *finitely definable* if there is a finite collection of symbols from L such that these symbols refer uniquely to the object when they are interpreted according to the dictionary D.

§6. Each sequence consisting of the numbers 0 and 1 only, including 0000 ... and 1111 ..., will be termed *binary sequence*.

Two binary sequences will be defined to be equal if and only if the entry in each place of the one sequence is the same as that in the corresponding place of the other sequence.

By a well known argument due to Cantor the totality of all binary sequences is uncountable.

That is to say, given any countable collection of binary sequences, there exists a binary sequence, which may be termed the *anti-diagonal sequence*, and which has in the n-th place an entry which is different from that of the n-th binary sequence of the countable collection. This binary sequence (the anti-diagonal one) cannot be contained in the given collection. Therefore no countable collection of binary sequences can contain all of them.

The set of all the sequences which are finitely definable using D must, by contrast, be countable.

One can impose an ordering upon the collection of all finite combinations of symbols of the system L, thereby constructing a countable sequence, in the following way: Those combinations that possess the fewest symbols appear first and among those with an equal number of symbols the "alphabetical" order is to hold. By means of this ordering all words that define a definite binary sequence, and hence all finitely definable binary sequences, are enumerated. One and the same binary sequence could appear on the list more than once – this does not matter.

§7. From these considerations it follows that there do exist binary sequences which are not finitely definable using D.

This is not surprising in itself, for, after the binary sequences have been defined in their totality, it would be a restriction to require in addition that every single binary sequence must possess its own separate finite definition.

The paradox lies, however, in the fact that from among those binary sequences which are not finitely definable a definite one can be unambiguously specified, namely: *the anti-diagonal sequence associated with the collection of finitely definable sequences*. This seems to give us a finitely definable sequence.

In reality this is not the case. This last definition certainly does consist of symbols from L and words which are in D. But it is for precisely this reason that it is logically objectionable.

That is to say, anything determined by means of an unambiguous definition consisting of words from D must be finitely definable. A binary sequence cannot however be finitely definable *and* at the same time meet the conditions of the antidiagonal sequence. That definition therefore requires something inpossible, and consequently there does not exist any binary sequence which satisfies it.

This result is to be expected. If something cannot be defined in a certain way, then all attempts to do so are bound to fail.

The given definition becomes meaningful when it is taken out of formal symbolism into pure thoughts where it is abstracted from its formal expression. Then it clearly defines a definite binary sequence which is not finitely definable using D. (This binary sequence does not even allow a purely formal representation by introducing a new symbol – cf. beginning of §4. After all, the formally represented concepts of the future are also all contained in D.)

§8. A simple example should clarify this situatuation (cf. Finsler [1925]).

The numbers 1, 2, 3, and the sentence "the smallest number which is not written on this board" are written on a blackboard. The quoted phrase which is written on the blackboard cannot define any natural number in a meaningful manner. One cannot, however, conclude from this that the number 4 does not exist. Equally one cannot conclude from the above paradox that binary sequences which are not finitely definable do not exist. The same holds true of real numbers which are not finitely definable.

III. A Proposition which is Formally Undecidable

§9. A proposition which is formally undecidable will now be constructed. In doing so we adopt the following definitions as a result of the previously established ideas.

A *formal proof* is a finite combination of symbols of the system L such that its meaning using D yields a logically correct proof.

Any proposition is said to be *formally undecidable* if no formal proof is possible, either for it or its negation.

We now consider all those combinations of symbols of the language L which furnish a formal proof for the fact that the number 0 occurs infinitely often in a given binary sequence, or alternatively prove that it does not occur infinitely often. Every such proof determines a unique binary sequence, the one treated in the proof. There could exist multiple proofs for one and the same binary sequence.

These proofs can be arranged in a countable list using the ordering defined in §6. This gives an induced enumeration of the corresponding binary sequences. Now take the anti-diagonal sequence determined by this induced list and consider the proposition:

In the anti-diagonal sequence so defined the number 0 *does not occur infinitely often.*

This proposition is formally undecidable since the binary sequence to which it refers cannot belong to the previously drawn up list. For this reason this proposition can be called *formally consistent.*

§10. In spite of this, however, one can see that this proposition is false and consequently inconsistent.

That is to say, the number 0 must indeed occur infinitely often in the anti-diagonal sequence under consideration. To see this, form increasingly long proofs of the proposition that the number zero does not occur infinitely often in the sequence whose entries are uniformly equal to one. To every such proof, however, there corresponds a zero in the anti-diagonal sequence. Hence infinitely many zeros must occur in this anti-diagonal sequence.

§11. It appears that there is a contradiction here: A formally undecidable proposition seems nevertheless to have been formally decided.

In actual fact this is not the case. The formal proof just given, which consists of words from D, is, as such, formally objectionable; for it refers to a binary sequence which cannot occur in the enumerated list. Hence this very proof implicitly requires of itself that it is not a formally valid one. Thus were we to suppose that the proof be formally valid, then a contradiction would appear in the proof itself; so it cannot be formally valid. There is no way around this difficulty that would allow one to formally decide the anti-diagonal proposition.

The proof becomes free from objection, however, as soon as it is transfered from the formal symbolism into pure thoughts where it is abstracted from its formal expression.

No contradiction is to be found within the formal definition of the anti-diagonal sequence itself; the sequence can be formally defined.

Thus there really does exist a formally representable proposition which is formally consistent but logically incorrect.

§12. A simpler example, like the one in §8, will be given by way of comparison.

One writes on the blackboard the proofs of the four propositions:

$\sqrt{1}$ is rational; $\sqrt{2}$ is irrational; $\sqrt{3}$ is irrational; $\sqrt{4}$ is rational.

Further one writes in addition the following on the same board:

"*Definition.* Let m be the smallest natural number such that no proof on this board decides the rationality of its square root.

Assertion. $\sqrt{m}$ is irrational.

Proof. Since for the numbers 1, 2, 3, 4, it is decided on this board whether their square roots are rational or irrational, m must be greater than 4. Since proofs for at most five numbers are on this board, m must certainly be less than 9. Therefore the square root of m must lie strictly between 2 and 3. Now the square root of a natural number is either a whole number or an irrational number. Since $\sqrt{m}$ cannot be whole number it follows that $\sqrt{m}$ is irrational."

In this example m must be equal to 5, because for $m > 5$ the number 5 would certainly be the smallest natural number for which nothing is proved on the board.

It really is the case that $m = 5$; for the last proof which stands on the board is mistaken just because it appears there. By the very fact that it is written on the board it is necessarily invalid; otherwise m would not be the correct number.

The proof given is correct, however, as soon as one grasps it purely in thought, so that as such it does not stand on the board.

The definition of m given on the board is sound; it clearly defines the number 5.

If, in addition, one were to write on the board the proposition: "$\sqrt{m}$ is rational", then this would be a proposition which would be represented on the board but not refuted nor even decided there. It would nevertheless be false. Further, it would be impossible to write on the board a proof which could refute this proposition.

First published as: "Über die Lösung von Paradoxien" *Philosophischer Anzeiger* 2 (1927), 183–203.

On the Solution of Paradoxes

The antinomies of set theory and similar associated paradoxes have frequently been the subject matter for discussions; but full agreement over the various explanations has never been reached. As these things are of special significance for the foundations of mathematics, it is a pleasure to take the opportunity offered by the publisher of this periodical to comment from a mathematical point of view on the paper of H. Lipps, *Die Paradoxien der Mengenlehre* [1923], in the *Jahrbuch für Philosophie und Phänomenologische Forschung*.

In this investigation Lipps was, as he says, aiming at a solution of the paradoxes with which set theory is encumbered and in doing so he adopted the standpoint that they need, and admit a solution. In my opinion, the mathematician must also adopt this point of view. Every inconsistency in pure mathematics or in logic imperils the entire body of scientific knowledge and must therefore be cleared up and done away with. This task can be difficult but surely not insoluble; for even if there do exist problems which cannot be treated by human means, the task of exposing mistakes in an obviously fallacious argument is surely not one of them. On the contrary, in this case we must be capable of carrying out a solution.

A great danger arises here, that of being content with the discovery of a genuine or even only imaginary mistake and then considering the question as being solved without ever facing the heart of the matter. The contradictions will then appear again at deeper levels. One ought to place more rigorous conditions on a genuine solution. Above all it must not carry arbitrary elements in it; on the contrary, it must arise unambiguously from the task itself. The solution of a contradiction can therefore not be found in philosophical discussions of a general sort, but only in logical reasoning in which the concepts are exact.

I should like to discuss only a few of the paradoxes by giving briefly the most important points on which I cannot agree with the author of the paper mentioned, and will add the explanation which I hold to be correct.

A word is said to be *autological* if its meaning applies to itself; it is said to be *heterological* otherwise. Thus the word "short" is autological; but the word "long" is heterological. Now, is the word "heterological" autological or heterological? Both assumptions lead to a contradiction.

Lipps seeks the explanation by saying that here we are not concerned with properties of words. But *why* should it not be a property of a word that its meaning applies to the word itself? If the assertions "the word 'short' is short" and "the word 'pentasyllabic' is pentasyllabic" possess a common property then the words "short" and "pentasyllabic" themselves also have a common property, namely that assertions of this kind hold for them.

In order to clarify this point we want to investigate what has to be understood by a "property". First of all, one recognizes that even arbitrary things which are not of the same kind can possess a common property. Thus Europeans and Sirius have the common property of being mentioned at the beginning of this sentence, and hence differentiate themselves from other "things". In general, in an arbitrary domain of things there are at least as many properties as there are sets of these things; for to every set there corresponds a property of its elements, namely that of belonging to this set. Together with further requirements, which have to be placed upon the concept "property", one comes to the following definition: Anything that applies to or does not apply to each object of a given domain is said to be a property. A property must, therefore, unambiguously subdivide the things of the domain into those to which it applies and those to which it does not. Conversely, anything which gives rise to a division of this kind is said to be a property. The partition, however, need only be determined for the domain of things in question; outside of this domain the property can be "meaningless" or "undefined".

Other definitions which could be given for the concept "property", if they are sufficiently exact, can be reduced to this definition. Intuitively, properties are often indeterminate, but they can be made precise, at least in principle, by refining the definition in the doubtful cases or by restricting the domain.

Thus if one speaks of a *definite* property, it must follow unambiguously from the definition to which things it applies and to which not. Nothing more, however, has to be required of the definition; for from the purely logical point of view, there is no obstacle to considering properties of equal extension (those that apply to the same things and fail to apply to the same things) as being identical. Among the positive even numbers the property of being "prime" is identical to the property of being "smaller than four", as well as to the property of being "equal to two". In the domain of all numbers, however, quite different properties are designated by the quoted words.

Now, how does it stand with the concepts "autological" and "heterological"? Are definite properties of words expressed by them or not? In the following investigation we must adhere strictly to the given definitions of the concepts, since they are determined solely by means of these definitions and do not carry a definite meaning in themselves.

Let us first of all consider the three words "short", "pentasyllabic" and "long". Of these, according to the definition, the first two are autological, the last heterological. In the domain consisting of these three words, the properties are well-defined.

For things other than words, single letters for instance, these properties are not defined. At first sight it seems as though these properties are defined in the domain of all words, and hence would have an unambiguous meaning for any arbitrary word. This, however, is not the case.

In order to know whether the meaning of a word applies to the word itself, one must already know the meaning of the word, not only in certain single cases but also as it applies to itself. Concerning the words "autological" and "heterological" the meaning, according to the definition, is dependent on this very meaning itself in a circular manner. This circular definition can indeed be fulfilled in the case of "autological" but not fulfilled uniquely. In the case of "heterological", however, it is completely non-satisfiable. Thus the definition does not yield an unambiguous result for arbitrary words to which it could apply. Exactly for this reason these properties do not have a definite meaning in the domain of all words. If, however, one wants to use the words "autological" and "heterological" to designate properties which are defined for these words themselves as well as for others, then a special stipulation is necessary. One could rule that both words are unambiguously heterological, without regard for the earlier rule. In this case the word "heterological" would indeed have the property that its meaning applies to itself, but it would not possess the property of being autological.

In summary, a property is only determined by its definition, and therefore only has meaning concerning those things for which the definition is unambiguous. That the definitions of "autological" and "heterological" do not apply to arbitrary words stems from a circle contained in their definition.

Thus if one says: It is necessary for a property either to apply or not apply to each thing, then this is a *stipulation* or a *requirement* which one places upon the concept "property" and not a self-evident *statement*. It is not necessary that this requirement be made universal. One can, as we have done, speak of properties when the domain is restricted.

Other paradoxes are now to be dealt with in just the same way. The predicates "conceivable" and "abstract" can both be stated of themselves; and this constitutes a property common to both predicates. If, however, one designates as *predicable* all predicates which can be stated of themselves, and on the other hand all others as *impredicable*, then two properties and at the same time two predicates are defined by this. But they cannot be applied to quite arbitrary predicates, as one might naively think. This is because there is a circle in the definition of the predicates "predicable" and "impredicable", so they do not always yield an unambiguous result. A further special condition is still necessary if the predicates are to be applied to themselves. There does not exist a property "impredicable" which applies to *all* predicates. One can indeed say that the predicate "impredicable" cannot be stated of itself, if it is defined as above, because when applied to itself the definition makes no sense. From this it does not, however, follow that "impredicable" would now in fact be impredicable. Suppose that "impredicable" were completely meaningful, then the definition would require that it apply to itself, but also would require that it does not apply to itself. Hence it remains meaningless and therefore cannot be applied to this predicate.

The "paradox of finite definability", too, stems from a circular definition. Every number is either finitely representable or it is not, in so far as we accept that the concept of "finite definability" is well-defined. The diagonal argument of Cantor gives a method for deriving, from the finitely definable numbers, a definite number which is not finitely definable. By means of the definition of this number, it does appear to be finitely represented after all.

This, however, is not the case, for the definition would require something impossible of the number which it purports to define, were it to be expressed finitely. Namely, on the one hand it would not belong to the finitely representable numbers: But on the other hand it would indeed be finitely represented, by means of the given representation. No number, however, can satisfy such a definition. If, however, the definition is grasped purely in thought, so that as such it is *not* presented "in finite representation", then it is free from objection and there is nothing that can be brought against the existence of the number in question (see Finsler [1926a]).

We may consider further the Russell paradox of the class of all classes which do not contain themselves, where one first assumes that every class is defined by means of a property possesed by its elements.

If the property of a class, through which its elements are determined, applies to itself then the class too must itself belong to its elements, i.e. it must contain itself. Now does there exist a class of all classes which do not contain themselves? The assumption of its existence leads to contradiction; for it could not contain itself, but again certainly would have to contain itself. On the other hand, however, "not to contain itself" is a definite property of a class and consequently a class certainly ought to be definable by means of this property.

The solution of this paradox cannot depend upon the questions: How are classes related to their elements? Are they associated concepts, or not? If each property really defines a class, then this must also hold for the specified property.

In order to find the solution we must investigate whehter it really is the case that every property defines a class. For this, however, it is necessary to know exactly the concepts "property" and "class". Indeed it is not a matter of knowing what these concepts mean "in themselves" but how they are defined. The concepts are fixed only through their definition and they also have meaning only in so far as the definition has meaning. If the concept given by means of the definition does not coincide with the concept which one wants, then there remains nothing more to do than to modify the definition and seek for a more suitable one. If, however, the concept wanted is logically impossible; that is, if one imposes mutually contradictory requirements on the concept, then one will not succeed in obtaining a suitable, logically unobjectionable definition for it.

Let us first of all assume that a definite domain of *concrete* things be given. In this domain a property E is defined as soon as it is determined for every single one of the given things whether E applies to it or not. We can now stipulate that to each such property there corresponds a class K, and that those things to which the property E applies are to be designated as elements of the class K. The domain of things under consideration is in no way altered by this stipulation and there are thus no difficulties here.

We can, however, no longer reason in this way when we want to consider arbitrary classes of classes. The concept of the class in the previous example was tied to the concept of property, and this was based upon a given domain of things. If, however, these things are themselves arbitrary classes then the concept of a class is already needed for their definition. Thus it must be defined by means of itself. Now suppose that we succeed in determining definitely the concept of a class by means of a circular definition, or what amounts to the same thing, through an implicit definition. Then this would only show the inadequacy of the earlier definition so that the concept of the propery would not in fact satisfy the previous requirement that to every property there corresponds a well-defined class. In other words it is not possible to determine the concept "class" in such a way that to arbitrarily given classes there always corresponds a definite class which contains them as elements. The proof of this is just the fact that there cannot exist a class whose elements are precisely the classes which do not contain themselves. If one wanted to define a *new* class by means of the property of being an arbitrary class which does not contain itself, then the attempt would contradict itself. Because, of course, one assumes that *all* classes are already present so that there cannot exist any more that are new.

That classes are not defined by means of properties, or cannot be so defined, stems from the fact that the properties still do not have any definite meaning as long as the domain of things to which they refer has not been fixed. So, in particular, the property of being a class which does not contain itself only has a meaning when the concept of a class has been fixed. If one defines the concept "class" without using the concept "property", then subsequently in the domain of all classes one can fix the concept "property" in the earlier way, and one then finds that there does not correspond a class to every property. A definite property does indeed give rise to a subdivision within the domain: the things to which it applies and things to which it does not apply. One surely is used to seeing classes defined in this way. Attention must, however, be paid to the fact that in this case one imagines things of the domain as something other than subdivisions or properties of these things; in some instances identification of the subdivisions or properties with the things themselves could turn out to be impossible.

If one wanted to determine the concepts "property" and "class" simultaneously in such a way that to every property there corresponds a class and to every subdivision of the classes there corresponds a property, then one would not succeed because of a non-satisfiable circle. Here attention must be paid to the fact that in certain cases circular definitions can indeed possess unique solutions, but that they need not be satisfiable in every case.

In summary, we can therefore say: The concept "class" (and similarly the concept "property") is not given in itself but is something which first of all has to be defined. Every attempt at defining the classes generally in such a way that they can also occur arbitrarily as elements, necessarily leads to a circular or implicit definition. It is impossible to determine the concept "class" consistently in such a way that arbitrary classes together always represent the elements of a definite class.

There still remains the question as to how the concept "class" can be appropriately defined. Not all requirements can be satisfied, so certain restrictions are necessary. In many cases one will be able to avoid the circle by admitting only those things as elements that are independent of the concept "class". Such a restriction is, however, too narrow, especially for the purpose of set theory, which has to do with analogous problems. Here, however, one can impose another restriction which allows one to investigate those peculiarities that are caused by the circle and to uniquely determine the concept of a class.

The difference between classes and the sets of mathematics lies essentially in the greater exactness of the latter. Lipps says that the existence of a class only presupposes the existence of *some* elements having a certain property but not necessarily the existence of *all* of them. It has to be remarked that either *all* the elements are determined when the property is specified, or the property (and also the class) will not have a well-defined meaning. Sets too are only *related*, though in a definite way, to their elements, without actually "consisting" of them; for, there is an essential difference between the *one* thing, the set, and its elements, which are usually *many* things. It is not necessarily excluded that a set contains itself, that is, that it be one of its own elements.

Now if one wants to define sets exactly for mathematical purposes, then one must also sharply circumscribe the domain of things which are permitted to occur as elements. Otherwise the concept of "all" sets does not have any definite meaning. The above restriction, which leads to an exact definition, consists in the fact that one admits only sets as elements; that is, one considers only sets of sets, without requiring that sets collected together arbitrarily always form a set. Through every set, however, its elements must be uniquely determined. If one adds a further, suitable stipulation which determines when two sets are to be considered as being identical, and then collects all sets which are possible in this way, then one obtains as can be shown, a unique and consistent system, onto which further investigations can be built (see Finsler [1926b] and [1925]). Genuine antinomies can no longer occur here; for a real contradiction in logic itself is not conceivable. Apparent contradictions can occur only through mistaken reasoning. In particular, the antinomy of the set of all ordinal numbers disappears as soon as one keeps to exact definitions. The more exact development, however, leads too far into the realm of pure mathematics.

First published as: "Gibt es unentscheidbare Sätze?" *Commentarii Mathematici Helvetici* **16** (1944), 310–320 (MR **6**, 197).

Are There Undecidable Propositions?

1. Formal Systems

Approximately 18 years ago I showed that in formal systems of a general kind one can specify propositions which are not decidable by means of formal proofs within the systems themselves, but which nevertheless can be decided by virtue of their conceptual content (see Finsler [1926a]). A formal proof was considered to be admissible for the purpose of this argument only if its interpretation constituted a logically unobjectionable proof.

Later Gödel [1931] attempted to specify similar formally undecidable propositions within the system of Principia Mathematica and other related systems. In order to do this he constructed a proposition which asserts its own formal unprovability. By considering the conceptual content of such a proposition one sees that it must be true when the system is consistent, thus it is really unprovable and therefore also undecidable. That is to say, were it to be false then it would have to be formally provable and consequently true. It is remarkable that some who have rejected informal reasoning in general accept such arguments as valid in this special metamathematical case.

Now for Gödel, however, a formal proof must satisfy certain merely formal restrictions without regard for its actual meaning. This notion of proof, customary in the study of formal theories, does not present any hinderance to an attempt to represent Gödel's argument within some formal system. Furthermore, the argument of Gödel for formal unprovability given above is conceptually correct. These facts, however, contradict the claim that a statement completely independent of formal provability has been found. It follows that Gödel has by no means demonstrated the existence of propositions which are formally undecidable in an absolute sense. But he has shown that the systems which he has taken into consideration are formally inconsistent if they permit certain simple implications which are conceptually correct.

If these systems are to be consistent then at very least one of these conclusions must be absent. Actually in the usual system (as P. Bernays has confirmed for me) the following implication does not occur:

"From the provability of A follows A",

or in symbols:

"Prf(A) → A".

One is thus not allowed to employ this implication within the framework of formal systems, at least not for proofs!

If, however, this implication were to be incorporated into the system then the concept of provability would change and indeed, as has already been remarked, to such an extent that either the formally undecidable proposition is no longer representable or else the system becomes inconsistent and consequently every proposition becomes provable. In order to avoid this one has to prohibit the incorporation of this implication.

It should be self-evident that a true proposition becomes "undecidable" when a method of argument necessary for its proof has been ruled out. If one forbids the use of the principle of complete induction then already $a + b = b + a$ is an "undecidable" proposition for the natural numbers.

The "inconsistency" of certain more recent formal systems has been demonstrated by S. C. Kleene and J. B. Rosser [1935]. This could very well give the impression that the contradictions in these systems stem from the inclusion of inherently inadmissible methods of argument. According to the foregoing remarks, however, this does not need to be the case. Inherently correct conclusions relating to the formalization, can become false through just this formalization, so that therefore the formalization does not lead to "greater exactitude" but to disintegration.

If, however, in considering proofs one observes not the form alone but, as was remarked at the beginning, pays attention to the inner meaning and contents then the contradictions disappear. One no longer needs to set up prohibitions and can in fact specify formally undecidable propositions. But if one attempts to represent the proofs for the truth of such propositions formally, then the attempt miscarries, because these proofs become inconsistent with respect to contents through just this formalization; thus they are false and therefore invalid. The proofs become false through their formal representation in the same way as the assertion "I am silent" becomes false the moment it is spoken out aloud.

As can be seen these things are very closely connected with the logical paradoxes; firm foundations cannot be found as long as no clarity exists concerning them. Now I have already stated repeatedly how the paradoxes are to be solved ([1925], [1927b], [1941b]); it appears, however, that this has hardly been noticed or understood. I will therefore discuss the paradox of the "liar" here.

The conviction that every paradox must be soluble puts one on the road to success also concerning propositions which assert their own unprovability in an absolute sense. In this way one obtains information concerning the existence, or non-existence of absolutely undecidable propositions.

2. The Paradox of the "Liar"

One could ask whether the paradoxes belong to philosophy rather than to mathematics. In philosophy many opinions can arise; by contrast, in mathematics only a rigorous, objective distinction between true and false is allowed to exist. This point of view can, and shall be adopted in connection with the treatment of the logical paradoxes; these are therefore to be counted as being a part of pure mathematics.

It is possible to specify propositions which assert their own falsity, for example the sentence "I lie", or the assertion "The assertion standing here is false". The question arises as to whether such a proposition is true, false, or perhaps even meaningless.

One reasons as follows: If the given proposition were to be true then it would have to be false, but if it were false then it would have to be true. Thus the proposition can be neither true nor false. Therefore it is meaningless, or has at very least "no well-defined meaning".

This line of thought, however, is not tenable, for it leads into a contradiction. We will define any proposition to be meaningless if it is neither unambiguously true nor unambiguously false. A true proposition is not meaningless: similarly, a false proposition is not meaningless.

Now, if the given proposition were meaningless then it asserts something that is unambiguously false; for the assertion states that the proposition is false, and thus not meaningless. So it follows that if the proposition were meaningless, then it would be false. And yet the proposition cannot possess a "variable" meaning, if only because of its invariability.

The right solution follows from the observation that, when considering an assertion, one has to pay attention not only to the linguistic expression but also to the real meaning. Every assertion has the meaning that what it asserts is true. If, however, it is asserted at the same time that just this meaning is to be false, then there are two contrary assertions and these, taken together, yield a false assertion. "A and not-A" is always false, quite irrespective of whether A or not-A is true. The given proposition is therefore unambiguously false. (A similar argument appears in Geulincz [1663].)

From the fact that the proposition is false it does not follow, however, that it must then nevertheless be true. For a proposition can very well be false even though part of what it states is true. In connection with this proposition one can distinguish between an explicit and an implicit meaning. The explicit meaning that the proposition is false would be true if taken on its own, but together with the implicit meaning, that it is true, a falsehood arises. If I say, however, that the assertion "I lie" is false, then this is a true proposition. Whether the proposition relates itself to itself or to another proposition constitutes an essential difference.

One of the main reasons for the emergence of the paradoxes is that the implicit statements, which really are present, are overlooked or not taken into account, and this danger will at all events only be intensified by a purely formal treatment.

Some writers object to considering the conceptual content, saying it is too "vague". They hold the opinion that only a purely formal representation can be sufficiently "precise". This striving to completely replace the meaning by means of formulas resembles an attempt to judge the colors of things solely by their form. It is understandable that people who are color-blind possess a great interest in such "formal" definitions; but that this is the best way in which to arrive at information concerning colors can very well be open to doubt.[1]

In the domain of conceptual content there is no distinction sharper, more precise than that between true and false. This is clear if one adheres to the law of the excluded middle, investigating only those things to which this law applies. Everything must be true or false: nothing can be both without destroying the whole. One also recognizes that every logical paradox must be soluble. A contradiction cannot arise out of nothing: it can only come out where it has already been put in. One thus only has to avoid contradicting oneself in order to maintain consistency in mathematics.

3. Absolute Decidability

From now on we will consider unambiguous propositions, that is to say, propositions that are either true or false. Every assertion

---

[1] See Tarski [1935]. From §1 of this work it emerges that he cannot decide which of the propositions introduced by him are actually true and which false.

which is not unambiguously true or unambiguously false will be said to be "meaningless".

Now consider the following proposition:[2]

"The assertion standing here is unprovable."

Take not only the formal proofs, but ideal proofs also, provided that they are conceptually correct. This last requirement means that from the provability of a proposition its truth must follow. This is a necessary condition which has to be placed upon the concept of provability. No further restrictions shall be imposed upon this concept. Otherwise it would be conceivable that a proposition which could not be proven under such restrictions, would turn out to be provable without them. We want to avoid this possibility. Thus we consider the concept of provability in its greatest possible scope.

It is plain that one cannot simply require that every true proposition be provable. On the contrary, one can pose the question: do true propositions exist which are not provable? These would be absolutely undecidable propositions.

If the proposition given above, which asserts its own unprovability, were to be true then it would be an example of such an undecidable proposition. Yet in order to know that it is true one would have to prove it and this is impossible according to what it states. So this is not a possible approach to solving the problem.

Let us investigate the proposition more closely. Right away one comes up against a paradox. That is to say, the proposition cannot be meaningless; for it would follow that the assertion carried by it and the proposition itself would be true. A meaningless proposition certainly is unprovable; for a provable proposition is true, and therefore not meaningless.

On the other hand, if one now makes the assumption that the proposition and the assertion carried by it are false then it would not be unprovable but provable, and consequently true. If, however, the proposition cannot be meaningless nor false, then it must be true. So we seem to have proven that the proposition is true. Yet, for an unprovable assertion, no such proof can exist.

In order to solve the paradox one must once again make allowance for the implicit statement of the proposition which requires that the assertion be true. It must now be investigated whether or not this assertion is compatible with the explicit statement that it is unprovable. With this we return to the earlier problem: do there exist true but unprovable propositions?

---

[2] A similar formulation is to be found in Hilbert/Bornays [1939, 269–270].

In the first instance let us make the assumption that such propositions cannot exist. Then it would follow that the two concepts "true" and "unprovable" are incompatible with one another. The implicit and explicit statements of the given proposition would contradict each other; the proposition itself would be false. It can also be seen that the proposition can be false only if the two concepts are incompatible with one another. Otherwise, the falsity of the proposition, the explicit assertion itself, would imply that the proposition is not unprovable but provable and consequently not false.

Let us now make the opposite assumption: that there do exist true but at the same time unprovable propositions. Then it follows that the two concepts, true and unprovable, are compatible with one another and the given proposition cannot be false. It has already been shown that the proposition can not be meaningless; so it follows that, under the assumption made, the proposition must be true.

We can now draw the following conclusion: If the absolute undecidability of some unambiguous proposition is provable then the last assumption above is also provable, i.e. it is then provable that there does exist at least one true but at the same time unprovable proposition. Either that proposition or its negation would be an example. This means, however, the proof just given for the truth of the proposition which asserts its own unprovability has been completed. This yields a contradiction. We thus have the following result: *There does not exist any unambiguous proposition for which the absolute undecidability is provable.*

There still remain two possibilities, namely: *The proposition: "There do exist unambiguous, but absolutely undecidable propositions" is either false, or it is true but unprovable.*

Were it provable, then one would have a proof that there do exist propositions which are true and at the same time unprovable. This, it has just been shown, leads to a contradiction.

In the first case, the proposition is false and it is then still possible to prove this. Whereas in the second case it is at best refutable. This can now be expressed as follows: *The assumption that there do not exist unambiguous, absolutely undecidable propositions, and thus that there do not exist any "insoluble mathematical problems", is not refutable; it is therefore "absolutely consistent".*

Hilbert [1926] posed the problem of showing that a corresponding assumption does not yield any "finite" contradiction; as can be seen the problem allows a solution in the absolute sense.

If one adopts the standpoint that every "absolutely consistent", i.e. non-refutable proposition, may be taken to be true then every false proposition would have to be refutable and consequently every true proposition provable; here the concepts "true" and "false" are seen as absolute opposites. The proposition which asserts its own unprovability is then necessarily false. It has still to be investigated, however, whether this standpoint is justified.

A contrary assumption would mean that: There does exist a false proposition which is not refutable. But what does it mean that a proposition is false? In mathematics it can have no other meaning than that it contains a contradiction.

Now in mathematics we make frequent use of the assumption that something existent can also be taken as being given. This is in any case a self-evident principle in so far as it concerns the abstract property of being given and not that of practical constructibility.[3]

Were a contradiction contained within a proposition to be given, then the proposition would be refuted by means of this very contradiction.

Thus it follows that every false proposition really is ideally refutable. Yet if the negation of a proposition is refutable then the proposition itself must be provable. So it follows further that every true proposition is provable and consequently *every arbitrary, unambiguous proposition is ideally decidable.*

The problem of ideal decidability can thus be solved very simply in this way, by leading everything back to the meaning of the concepts; the previous considerations are not superfluous however: they touch upon other conclusions which are also significant for practical decidability.

One may raise perhaps the following objection: It is conceivable that a poposition A contains no contradiction and that the negation of A, that is not-A, is likewise consistent. It is then only with the "conjunction" of both, "A and not-A", that a contradiction arises. How is the decision then to be met?

In this case the proposition A cannot be false, but as not-A is also not false, A cannot be true. Thus it follows that A is meaningless. The conjunction of a meaningless proposition with its negation still yields a contradiction.

On the other hand the assumption that A is meaningless does not yield a contradiction. That is to say, if A were not meaningless then it would follow that A would have to be either true or false. This, however, would then mean that either A or not-A contains a contradiction which is, by hypothesis, not the case.

But does not a meaningless assertion have to be false, since it implicitly asserts, contrary to fact, that it is true?

---

[3] This principle applies to set theory though it is not related to the axiom of choice.

For propositions which really represent assertions this interpretation is in fact correct; but then there would not exist any meaningless assertions, only ones that are true or ones which are false and so the case just considered could not occur. Actually, if the proposition A represents a genuine assertion and does not contain a contradiction then the negation of A asserts that, in spite of this, A is false. The negation thus contains a contradiction and is consequently false. The negation of A always states: A is false. The whole proposition must be negated and not merely the explicit statement contained in it. It contradicts the concept "false" to say of a consistent assertion that it is false.

On the other hand it is possible to think of propositions which do not have the meaning that their unrestricted truth ought to be asserted. According to our determination such statements are to be designated as meaningless, even though they may have a meaning in other contexts.

The parallel axiom could be taken as an example, or even the proposition:

"The number n is an even number."

In this example everything hinges upon the meaning given to the sentence. If nothing further is stated concerning the number n, and the proposition is meant as an unrestricted assertion, then it is false; for the number n could equally well be an odd number. If, however, the proposition signifies only an assumption, a postulate, condition, or stipulation then it is not an assertion. As an assertion it is meaningless, even though it can be meaningful as an assumption.

4. Practical Decidability

The results obtained so far still do not supply a means whereby a decision can actually be carried out for an arbitrary proposition. However, they are certainly of more than mere theoretical significance. In particular they can act as a stimulus (see Hilbert [1900]) in approaching difficult problems and could facilitate their solution. After all, it is certainly much easier to seek and find something when one knows that it is there than when one has to reckon with the possibility that it may not exist at all.

The following inference is of further practical significance: *For any unambiguous proposition whatever, that is for one which is either true or false, it is impossible to prove its absolute undecidability.*

Thus, Cantor's continuum problem can by no means be disposed of by showing that the conjecture concerned is neither provable nor refutable.

Further one finds that: *If it can be shown for an unambiguous proposition that it is not refutable in any way, then with this it has already been proved.*

As soon as one restricts the concept of provability these results immediately become modified. One obtains what at first seems to be a very difficult paradox which does indeed admit ideal, mental proofs that can be carried out by human beings. In order to make the matter clear I will restrict provability to the means I, myself, have available.

Consider the proposition:

"I cannot prove the assertion standing here."

Is this proposition true, false or perhaps even meaningless? The earlier results all break down as I can certainly not decide every proposition. However, the proposition given cannot be meaningless for otherwise it would definitely be true. The implicit assertion, that the proposition is true, does not stand in contradiction to the explicit assertion. And from the assumption that the proposition is false it follows that the explicit assertion taken for itself is false, i.e. it follows that I can prove the proposition. But a provable proposition must be true. With this I seem to have proven that the proposition must be true, and this can certainly not be the case; for if it is true then I cannot prove it.

The observation that my personal capabilities for accomplishing the task are not sharply defined yet play an important role here cannot in itself resolve the paradox. The question as to whether the proposition really is true or false remains.

In order to find the solution it is good to orient oneself by means of a similar paradox. As has already been shown above, if a proposition which asserts its own formal unprovability is true, then the corresponding proof can only be free from objection if it is not carried out in formal representation. Now here the case is similar: The proposition given above is true; however, the corresponding proof can only be correct as long as I do not carry it out myself. As soon as I attempt to prove the proposition I become entangled in contradictions and the proof becomes false. Thus it really is impossible for me to prove the given proposition. That I cannot prove it is of course something which I can establish; for, from the assumption that I could prove it, there immediately follows a contradiction. With this, however, I have by no means proved that

the proposition is true, for were it false then I would still be unable
to prove it.

May I, in spite of this, assert that the proposition is true?

Answer: Yes, but only because I have to believe the proposition. I
know that the proof for the proposition is correct as soon as someone
other than myself carries it out. Others can very well accomplish
what I cannot; this proof, however, is something that I cannot carry
out. Anyone else who wants to prove the proposition which states
that I cannot prove it does not become entangled in contradiction. If I
know, however, that others can prove a proposition which I cannot
prove then there remains nothing more for me to do than to believe
this proposition.

That there do exist things which one cannot prove but which one
must nevertheless believe has often been asserted. It is remarkable,
however, that here we are concerned with a definite statement for
which this can be established.

There still remains the question as to whether the foregoing
results are not strongly restricted by this last result, as we cannot
shed our human imperfections. Actually, it is highly probable that
one can form propositions of such a kind that their proofs turn out to
be so complicated in practice that it becomes impossible to succeed in
bringing them to completion (cf. Cérésole [1915]). The foregoing
considerations show, however, that propositions which we
demonstrably cannot decide must in some way or other refer to our
capabilities for accomplishing this task, and that thus they do not
belong to pure mathematics. This holds true also in the case in which
one could show directly that it is not possible for us to carry out
certain definite and in themselves perfectly possible mathematical
operations, nor reduce them to performable operations. Seen
mathematically these would only constitute an assumption and not a
proof.

If, on the other hand, one restricts oneself to purely formal
representations then one really is tied to an actually countable
domain, and thereby much that is of great value is lost.

73

Contribution to a discussion within *Dialectica* on the subject: "For and against Platonism in Mathematics – an Exchange of Thoughts". First published as: "Der platonische Standpunkt in der Mathematik" (Einem Vertreter der klassischen Mathematik, Herrn R. Nevanlinna zum 60. Geburtstag gewidmet). *Dialectica* **10** (1956), 250–255.

The Platonistic Standpoint in Mathematics

In Wittenberg's interesting paper *Über adäquate Problemstellungen in der Mathematischen Grundlagenforschung* [1953] and in the subsequent discussion (Wittenberg et al. [1954]) there is a clear description of a view of mathematics which is called naive or uncritical Platonism, "inhaltlich", theological, Platonic philosophy, classical philosophy, and by E. Specker [1954] the "an sich" philosophy. It is said, for example, in Wittenberg's answer that these views embrace the opinion "that this philosophy seizes upon *objective relationships*, that in itself it describes a factual situation, which as such remains removed from our powers of discretion". This "Platonic standpoint" is, however, rejected even if with regret. Why? First of all because of the antinomies!

Does one always have to be frightened out of one's wits in this way by the antinomies? Do people still believe in ghosts? Is it still thought that a contradiction can emerge anywhere, in places where it has not already been put in? If the antinomies really were to make the Platonic conception impossible, then one would have to be able to give at least one antinomy which cannot be explained in a reasonable way from a Platonistic point of view. *Which antinomy is it*? An exact and thoughtful answer to this question would interest me very much; no such antinomy has been known to me for decades.

Bernays does, to be sure, ask how far the set theoretic antinomies necessarily result from Platonism and holds it to be probable that there is a form of the Platonic view that does not lead to the antinomies. But, this does not require "an extraordinarily radical revision" of this philosophy, as Wittenberg expresses it, and in my opinion one does not even have to seek out this revised form; it is much simpler than that. The only naivety that has to be abandoned is the belief that many sets can always be formed into a single set. If one straight away considers many things as being one thing, then it is no surprise that contradictions can be derived. There is an essential difference between sets that contain many elements and those that have but one. But also the assumption that to many sets there must always exist one set which contains exactly these as its elements is objectively false, as examples show. This must not be postulated in any way whatever, whether our philosophy is Platonic or not. It is precisely in set theory that the relationships really are such that this assumption is not satisfied, and there is no cogent reason for holding on to this assumption.

Otherwise, I am, on the whole, thoroughly in accord with Wittenberg's critical remarks and should like, only for the sake of clarity, to mention a few points where differences arise.

Thus, after what has just been explained, one will not be able to say that the antinomies result as "an inevitable consequence of the naive Platonic standpoint quoted"; one could say "as being an *apparently* inevitable consequence". Further, the assertion that the set of numbers of the second transfinite number class does indeed exist, but not the set of all transfinite ordinal numbers, is just as little of a dogmatic character as is the assertion that 17 is a prime number but 15 not. These are simply facts. That in the one case one can perceive these facts a little more easily than in the other is of no importance. If it were true that we have at our disposal no criteria of any sort whatever for answering such questions, why is it then that this is not also true for the whole of set theory? Why is it that absolute anarchy does not rule throughout set theory? Why is it that someone who is convinced of the non-existence of an actual infinity can go to work using transfinite ordinals? Mere criteria for constructibility do not suffice in set theory; rules of construction could be rendered incapable of being carried out and thus lead to contradictions. But there are the criteria of truth and consistency: Are they to count for nothing?

When Bernays remarks that mathematical results are in no way put into permanent suspense because of the controversies surrounding the foundations, one can very well agree. But once again, how is this so? The reason is, of course, that in mathematical research one knows objectively what is true and false, without any special criteria and without codification. Why should this knowledge not be applied to foundational studies, set theory in particular? Every codification surely is a form of dogmatism as long as it is not supported by genuine insight.

In elementary calculation there is a prohibition against dividing by zero. This is not a dogma nor an arbitrary restriction but a well justified rule whose violation can easily lead to mistakes and thus to contradictions.

In the development of calculus, naive operations with infinitesimally small quantities proved to be very useful. In spite of this, it came to be rejected, because it became clear that these quantities, with the properties desired of them, could not be said to exist; they had no unobjectionable definition.

Similary, naive calculation with imaginary numbers proved very useful. In this case, however, these calculations could be retained, in spite of all sorts of doubts, because it was shown that the imaginary numbers can be defined consistently, and thus they really do exist. It suffices that the definition of the numbers be consistent; then one need not fear that a contradiction will arise in the course of a long computation.

Naive operations with sets have lead to contradictions and must therefore be reformed. It suffices to define sets consistently but without arbitrary restriction, and one then no longer needs to fear any antinomies.

In spite of these various constructions, the pursuit of antinomies seems to me to resemble the attempts at circle-squaring. A genuine antinomy would certainly mean that something exists and at the same time does not exist, and this is certainly absurd. By using false inferences one easily attains the goal in both cases. The only question is whether or not the errors are recognized.

Specker gives an instructive example in his inaugural lecture [1954]. He considers the set Q of those sets which do not contain themselves and are elements of another set; he then suggests that no contradiction arises with this definition. Accepting Q, however, would mean that there is no universal set. The assumption that the universal set exists, "for itself alone", is also free of contradiction. It is then deduced that from the consistency of the assumption of the existence of Q the existence of the universal set cannot be deduced.

It appears here as though one could consider either the set Q or the universal set as existing, according to one's choice, but not have both at the same time. "Existence", however, ought not to mean the membership of a set in some model of an artificially restricted set theory but its existence in an absolute sense. This means that it exists within the domain of all possible sets. The "possible" sets are, however, the ones which are consistent, that is precisely those for which the assumption of existence does not contain a contradiction. The existence of a set cannot depend upon our arbitrary choices, and thus cannot depend upon whether we assume some other set as existing or not. Therefore the question arises as to whether the set Q or the universal set exists.

The universal set is defined through the fact that every set belongs to it. From this requirement it cannot be concluded that some set does not belong to the universal set; this certainly ought to be very clear. This means, however, that the definition of the universal set does not contain a contradiction as long as the concept of a set is itself free from contradiction, and every consistently defined set is accepted. To exclude the universal set would be arbitrary. In the domain of all possible sets, therefore, the universal set must exist.

It is very different with the set Q. Since every set is an element of the universal set, Q is merely the "set of all sets which do not contain themselves"; this is not consistently defined and therefore does not exist.

It is not true that the assumption of the set Q "in itself alone" is without contradiction. If all men in a certain region of the earth are black, then it does not follow from this that the assertion "all men are black" cannot be contradicted. Since there do exist men who are not black, the assertion made is false, and thus inconsistent. Similarly, there can indeed exist a restricted model of set theory in which there does not occur a universal set, it may have Q instead. From this it does not follow that the definition of Q alone is consistent. On the contrary, it stands in contradiction to the general definition of a set. I do not see how one can honestly assert the existence of the set Q or reject that of the universal set.

I have given a consistent definition for sets in "The Infinity of the Number Line" [1954]. Objections against this have been raised by Kreisel [1954]. These rest upon the misunderstanding that set theory has to be a formalism instead of a theory ordered according to its meaning. Thus, it is not true that all the operations of set theory as they are to be found, for example, in Ackermann [1937], are permitted. In the domain of the circle-free sets these operations can be carried out, and are thus permitted, but in the domain of all sets they are not always possible and therefore are also not permitted. The axiom of choice, as well, is satisfied for circle-free sets [1926b, §17], but not for arbitrary sets [1941b]. The existence of the second number class follows quite analoguously to that of the sequence of natural numbers [1954].

When one reads further in Kreisel: "But for general M and N, $M = M'$, $N = N'$ may be separately consistent, though $M' = N'$ is refutable", then in any event $M = N$ is intended. If there do exist "logics" in which the relation of identity derived from equivalence is not transitive, then these could not be very useful for an exact knowledge. In a logic that recognizes meaning one has that two sets are always identical whenever possible, this means that the sets M and N are identical, provided they possess exactly the same properties. That is to say they are the same if the assumption that they possess the same properties does not contain a contradiction. Given that $M = N$, $M = M'$, $N = N'$ it then follows that $M' = N'$ holds. Consistency is to be understood in an absolute rather than formal sense, and therefore a contradiction cannot suddenly result from this.

With respect to the existence of the number line Lorenzen [1955] is of the opinion that many of the details in the carrying out of the proof do not appear convincing to him. It would be of interest to learn the reason for this, even for one of the details.

78

Contribution to a discussion in *Dialectica* on the subject: "For and against Platonism in Mathematics - an Exchange of Thoughts". First published as: "Und doch Platonismus" *Dialectica* 10 (1956), 266–270.

Platonism After All

It is particularly pleasing to see it clearly stated, in Wittenberg's "Why No Platonism?" [1956], that mathematicians would like to be Platonists, if only they were able. This wish can really be fulfilled, however, because the objections against it are not valid.

It is not a relevant objection merely to state that there are many mathematicians at present who do not accept the Platonist point of view. Hundreds of authors against one are not necessarily in the right. A mathematician, conscious of his responsibility, will not declare a proposition, for which he must be answerable, to be true just because many others do so, but only because he has convinced himself of its truth through his own deliberations. The truth of a proposition or theory ought indeed to lead ultimately to a "consensus"; the converse, however, need not hold! If there is no unanimity concerning the antinomies yet, then this signifies only that these things have still not been pursued with adequate understanding.

That there does exist exact mathematics, which includes the infinite, and that we can know it, may very well be considered as being a miracle, in just the same way as it is also a miracle that there do exist elephants, and that we can recognize and describe them. One difference between mathematical facts and those of natural science consists, however, in the fact that the former are of a compelling nature, that is, they simply cannot be other than they are, whereas the latter are of a more accidental nature. They could very well be different than we encounter them here. Natural science depends upon observation, whereas in mathematics only consistency decides. One could also very well say that mathematics is "necessity in thought". Yet, this should not lead to the notion that mathematical truths are dependent upon our thinking and therefore on the state of our brains. If this were so, then it would belong to the natural sciences. Mathematical facts are, however, independent of time, thus they are eternal, whereas the occurrences of nature are transient.

According to the conception put forth here, the "existence" of a mathematical object has a unique meaning corresponding to its consistency. There is no rational number whose square equals 2, and, in exactly the same way, there is no set which possesses exactly the ordinal numbers as its elements.

Now, an antinomy would result if, as Wittenberg [1956, 258] believes, we were to see ourselves being compelled "to 'form' mathematical objects (for example, the set of all transfinite ordinal numbers), that is, to *affirm* their existence". This ideal confuses the *class* of the ordinal numbers and the differently defined *set*. In an exact set theory sets do not arise through an act of collecting, but are mathematical objects with definite properties, in exactly the same way as the natural numbers are mathematical objects with definite properties.

Thirty years ago, I indicated the difference between sets and classes [1926b]. At that time this was characterized in a review (Skolem [1926]) as a "word game". I can in no way agree with this or with other aspects of the review. Today the distinction between sets and classes is commonly made. Wittenberg's arguments show clearly that people even now do not make the real difference clear to themselves.

If one respects this distinction and pays attention to it, then one can freely collect sets together into a class. There is no more reason to suppose that a class of sets is a set than there is to suppose that a set of natural numbers is a natural number, or that a flock of sheep is a sheep. The sets so defined as mathematical objects suffice completely for the whole of mathematics and do not lead to contradictions of any sort.

Now, however, in spite of this, in order to "save" the antinomies, one could require that the classes are themselves "objects" which can be collected together to produce objects of a new kind. Not only is this potentially dangerous, but it also has nothing to do with mathematics proper. Yet since this mode of argument is often brought forward it must be explored.

It is quite right that such considerations should not lead to a contradiction. This can only hold, however, when one does not impose unreasonable conditions.

If certain things are given, such as natural numbers or sets, then, as has already been observed, one can directly consider them, or an arbitrary selection of them, as a class. This does not mean, however, that one is allowed to iterate this process arbitrarily, by considering the class itself as being a single thing. If one does want to iterate the process (it is really unnecessary to do so), then one has to ask oneself how far it is permissible "to consider many things as one thing". One must also watch that in doing so one does not become entangled in contradictions or non-satisfiable circles.

Many things and one thing are not the same; that would plainly constitute a contradiction in itself. In addition, many things "considered as one thing" is something different from the same things "considered as many things". As has been remarked earlier, whether a set possesses many things or only one thing as an element constitutes an essential difference. If one does not pay attention to such differences, then it it clear that one approaches antinomies.

If one wants, however, to "collect together" many things into one thing, then this is an operation which in some instances cannot be carried out, because of a circle. To collect together exactly those collections which do not contain themselves is impossible. This is not an antinomy, but a fact. Now, one may perhaps say that the collection does not have to be formed, but is already there. One must keep in mind that something which is permissible in simple cases need not hold in general. The collection referred to is not already there; this really is so. It cannot be formed: the assumption of its existence contains a contradiction. If one thinks that the collection is formed simply through the fact that one speaks of all those collections mentioned, then this makes just as little sense as if one were to think that a greatest natural number is already formed by merely speaking of it.

When Wittenberg [1956, 261] says at the end of his remarks "that here fundamental demands which we are accustomed to place upon our thinking are at stake", this cannot very well mean that from now on contradictions ought to be allowed in mathematics, but only that one is not allowed to impose non-satisfiable conditions. Thus, in particular, one is not allowed to require that an operation must be performed even when this is impossible because of circularity.

To the points brought forward by Bernays [1956] I should like to make the following remarks:

1. In order to be able to know what is true and what is false in set theory one has to have won back the certainty lost through the antinomies. After clarification of the antinomies, however, nothing stands in the way. This does not mean that one could then solve every single question; this is not the case elsewhere in mathematics either. The form of the solution is, however, in my opinion, uniquely determined, though differences of opinion still arise concerning this.

With respect to the set of all sets there is certainly a great difference whether in the wording of a definition no direct contradiction is evident, or whether, as is the case here, it is directly evident that the definition does not contain a contradiction. The sets which are consistently definable are within the system of all sets. With this the question as to the factual situation is settled.

2. By "Platonism" one can of course understand various things. In the case before us this expression only means that in the realm of

set theory too, objective relationships do exist; it is not meant that sets would have to be given to us in some way other than our knowledge of their existence.

3. Consequently, quite another question is how we know these objective relationships and how are we able to base classical mathematics on them? Here we depart from the usual methods.

That the concept of the circle-free sets, indispensable for a proper founding of mathematics, has found a certain amount of attention after thirty years is also very pleasing. What is called "made plain by means of formalization" (Bernays [1956, 264]) refers to Ackermann's [1956] deduction of a formula corresponding to the axiom of infinity, using specific formulas which refer to the property of being a circle-free set. The actual existence of infinitely many things cannot be guaranteed in this way. This turns mathematics into a "doing as if", pretending there are infinitely many things. I cannot accept this.

II. Foundational Part

Introduction

The earliest of Finsler's set theoretic papers, *On the Foundations of Set Theory*, [1926b] remains the most important. The original title included the words: *Part I. Sets and their Axioms*; he expected to develop the theory further in a second part. When Part II finally appeared in [1964] it contained replies to some of Finsler's critics rather than positive developments of the theory. Part II is the last paper in this section, they have been ordered chronologically, but it is also of less mathematical importance than the others. Large parts of it are polemical; these parts reveal a dramatic clash between formalism and Platonism that bring a better understanding of the history of the foundational crisis. This relatively long paper also contains scattered but interesting, positive developments of the set theory presented in Part I. Before undertaking Part II one should see section VI below, of this introduction, for some background remarks on the axiom of completeness.

We believe, however, that most of the mathematical contents originally planned for Part II were eventually published in the other papers that are translated here.

Of these other articles *Concerning a Discussion on the Foundations of Mathematics* [1941b] is the most elementary because it presupposes only a knowledge of Finsler's axioms and does not concern the more difficult topic of circle-free sets. It provides some interesting examples to supplement *On the Foundations of Set Theory, Part I*, including a treatment of the Burali-Forti paradox and a refutation of the global axiom of choice.

Many readers may find it helpful to begin *Concerning a Discussion* even before finishing *On the Foundations of Set Theory*. The discussion of circularity at the end of *On the Foundations* is often found to be difficult and one could very well delay it and profitably pursue *Concerning a Discussion* first.

The definition of circle-free sets is the most fascinating, difficult, fertile, and controversial idea to be found in *On the Foundations of Set Theory, Part I*. The two articles, *The Existence of the Natural Numbers and the Continuum* [1933] and *The Infinity of the Number Line* [1954] cannot be fully understood without having first wrestled with Finsler's definition of circular and circle-free sets. They are not otherwise difficult to read however. These two papers are very similar. *The Infinity of the Number Line* is the report of a lecture; it is the more relaxed of the two articles. The subject matter is so remarkable that it seemed best to provide translations of both.

The issues raised in *The Infinity of the Number Line* are still as fresh as they were in 1954; this paper will surely repay the effort required to grasp Finsler's intention in defining the circle-free sets. The entire enterprise of producing simple, intuitive axioms, then defining the concept of circular and circle-free sets, and finally deriving the axiom of infinity, is truly astounding. This argument is destined to be of philosophical interest regardless of whether mathematicians take further interest in it. A book by Edward Nelson [1986], for example, maintains that the axiom of infinity is a mathematical superstition.

"Faith", Nelson writes in [1980, 80], "in a hypothetical entity of our own devising, to which are ascribed attributes of necessary existence and infinite magnitude is idolatry".

How can one reply to such finitist views? In ordinary mathematical practice one simply does not bother much with them. But the times demand that we do more than merely dismiss them on aesthetic grounds; there must be a rational treatment. Rational grounds for the acceptance of an infinite set are clearly too important to ignore.

In this connection the concluding remarks *of The Infinity of the Number Line* are memorable: "Think of the infinite as being locked up. If we want to obtain it, then we have to unlock it. In order to do this we need a key, and this must be turned. This turning is circular in nature. If no satisfiable circle is allowed, we cannot obtain the infinite: Should it be allowed, the infinite is obtained".

The remaining sections of this introduction are like footnotes to topics which may present difficulties or which have arisen in the mathematical literature. They have headings: beta relation, classes, the first axiom, and so on, so that one can turn directly to a topic that is of interest. The final section concerns Ackermann's set theory.

I. Beta Relation

Finsler always wrote the relation converse to set membership. In doing this he emphasized sets as axiomatically given objects rather than collections as in a naive theory.

II. Classes

It is taken as intuitively clear that one can form collections of given objects: Classes are collections of this kind. To collect things in this way is part of the basic logic that is indispensible to science. When we treat collections as objects, however, capable themselves of being elements of complex collections, we are thinking in a circle. These complex collections, the pure sets, must be introduced axiomatically: being essentially circular they cannot be understood in terms of something else. In this way we arrive at the need to axiomatize set theory.

III. Non-well-founded Sets

The usual sets of Zermelo-Fraenkel set theory are regular (or well-founded) as required by the axiom of regularity (also called the axiom of foundation). Non-well-founded sets were introduced by D. Mirimanoff [1917a], [1917b], [1920]. They are permitted within the Finsler theory. After all, in this set theory consistency is the standard by which we recognize existence.

Well-founded sets must not be members of themselves, $x \in x$; members of members, $x \in y \in x$; nor, in general, can they be part of a cyclic membership chain, $x \in y \in z \in ... \in x$. In addition, a well-founded set cannot participate in an infinite membership chain:

$$... \in x_3 \in x_2 \in x_1 \in x_0.$$

In other words, well-founded sets are those having only finitely long membership chains within themselves.

IV. The First Axiom

Finsler's Axiom I is: For arbitrary sets M and N it is always uniquely determined whether M possesses the relation to N, or not. It has been called the axiom of "definiteness". We usually think of it as insisting that sets be "well-defined". Some have written of sets being "sharply" defined. In any case, these sets are not "fuzzy".

To understand the axiom it is helpful to consider classes which fail to satsify its demands. The class of sets used to carry out Russell's paradox is a classic example.

Let $x \in R$ if and only if $x \notin x$. There is no harm in writing

$$R = \{ x : x \notin x \}$$

provided we remember that a mere class is given by these braces.

Russell's paradox is an argument which proves that R cannot be a set. If it were a set it would have to be well-defined, in the sense of Axiom I. We would have to be capable of determining in principle whether or not $R \in R$. But neither $R \in R$ nor $R \notin R$ is possible. Hence R cannot be a set.

The Burali-Forti paradox is a proof that the class of all ordinals is not a set.

Cantor's argument that there is no largest cardinal can be used in the Finsler theory to show, not that there is no universal set, but that certain subclasses of the set of all sets can not be formed into sets.

V. The Axiom of Identity

The second axiom tells us the circumstances under which two sets are to be recognized as identical. This axiom is subsidiary to Axiom I and applies only to things that meet its test.

The need for such an axiom can be easily seen by observing that otherwise the sets $J = \{J\}$ and $K = \{K\}$ could be taken as distinct.

There is a complication concerning this axiom: The original version contained an minor error which seems to have been mentioned by Finsler in [1928], so that the paper as it appears here still possesses a minor omission. The difficulty was independently detected by Peter Aczel in [1988] and he supplied a version of the axiom that seems to fix everything with the minimum disruption to Finsler's basic point of view. This error and Aczel's correction are described under the heading "the first version" below. This correction only is needed when one attempts to put section 7 of "On the Foundations" into modern notation, perhaps for the purpose of combinatorial study. One should certainly examine Finsler's paper before concerning oneself with the matter. The paper is not misleading as it stands; an awareness of Aczel's correction is unlikely to help the beginning reader.

Complicating the whole matter is the fact that Finsler later made a significant change in the Axiom of Identity.

We shall say that there are two versions of the axiom. The first version concerns isomorphic sets. It has the character of graph theory. Any such method of stating the Axiom, Finsler's original one, Aczel's modification and so on, will be considered as belonging to the first version of the axiom. The second version involves consistency and is without graph theoretic associations.

Finsler's First Version of the Axiom of Identity

In this section the isomorphism needed for Axiom II will be described in modern terms.

The transitive closure of x is written "Σ_x" in Finsler's notation. Let us also call it "$TC(x)$" so that we have available a suggestive notation that requires no subscripts. The class $TC(x)$ consists of elements of x, elements of elements of x, and so on. Finsler defines it as the intersection of the transitive sets that contain the elements of x. Literally x itself is not in $TC(x) = \Sigma_x$.

The fundamental idea of Axiom II, that isomorphic sets be identified, might be written as follows: $x = y$ if

(*) $\langle \, TC(x), \, \in \, \rangle \cong \langle \, TC(y), \, \in \, \rangle.$

This use of model theoretic notation would have been out of place in 1926. Its use, however, reveals a difficulty. To explain the matter, it is helpful to use graph theoretic vocabulary. The elements of $TC(x)$ can be taken as nodes of a directed graph having arrows directed from sets to their elements.

We have observed that x is not actually in $TC(x)$. When x is an ordinary, well-founded set, the graph in (*) would become disconnected, a disturbing though not necessarily incorrect state of affairs.

When one considers non-well-founded sets, one finds, by an example given in section 8 of "On the Foundations" that (*) is actually unsatisfactory. Consider the pair of non-well-founded sets $A = \{A, B\}$ and $B = \{A\}$. These sets will sometimes be called Fibonacci sets because of an analogy between A and B, on one hand, and the mature and immature states of Fibonacci rabbits respectively.

One has that $TC(A) = TC(B) = A$. Thus we have that:

$$\langle \, TC(A), \, \in \, \rangle \cong \langle \, TC(B), \, \in \, \rangle.$$

But A has two elements and B has but one. We wish to accept these as two distinct sets; so we must have $A = B$. This means that (*) will not do as an account of the isomorphism required by the first version of the Axiom of Identity.

Let us now use the term "transitive hull of x" to refer to the transitive closure of x, with x itself adjoined. Abbreviate it $TH(x) = TC(x) \cup \{x\}$.

Now, let us modify (*) so that

(**) $\langle \, TH(x), \, \in \, , x \, \rangle \cong \langle \, TH(y), \, \in \, , y \, \rangle$

becomes the condition for making $x = y$.

In this form we are dealing with rooted, directed graphs. Returning to the Fibonacci sets A and B which were used previously as an example, we could not have $A = B$ by the standards of (**), unless the isomorphism (**) were to match the two distinguished nodes, A and B, with each other. This is impossible since A has two elements (immediate descendents in the graph) whereas B has but one.

The explanations given here, of (*) and (**), were known to Finsler; we understand (**) to be the notion of isomorphism used in "On the Foundations of Set Theory", [1926b].

To see that a difficulty remains, one first of all forms the set $J = \{J\}$. Now consider the class D given by: $D = \{J\}$. Surely D is a set by the standards of Axiom I. Intuitively, $D = J$. In fact, this is required by the property of extensionality, proposition 5. But (**) alone does not literally permit this conclusion. The transitive hull

$$TH(J) = TC(J) \cup \{J\} = \{J\} = J$$

has a single element; but the hull of the hypothetical set $TH(D) = \{D, J\}$ has two. They cannot be isomorphic.

Strangely, (*) works well and gives $D = J$. From these facts we may judge that the proper account of isomorphic sets is a combination of (*) and (**) together.

Define a new hull, $T(x)$ of x as follows. If x is not in its own transitive closure (not "essential in itself" to use Finsler's term), put $T(x) = TH(x)$. If x is in $T(x)$, then introduce a new object X. Replace the x that is the root of the graph $TH(x)$ by the X; no other changes are to be made. In this case, x will still appear somewhere in $TH(x)$ since it is essential in itself. Now say that

(***) $\langle\, T(x), \in \,\rangle \cong \langle\, T(y), \in \,\rangle$

implies that $x = y$.

This new version (***) is the proper form of the axiom of identity. We do not need to introduce the roots of these graphs into the structural notation of (***) as was done in (**): An isomorphism of the graphs will necessarily carry one root into another. We have insured, through the definition of $T(x)$ that the roots of these graphs do not appear on the receiving end of any arrows in the directed graph.

This explanation of Axiom II can be used to make it properly precise for combinatorial studies.

The account here is like that of Aczel [1988] who also supplied his own, original criterion of set identity that is stronger than the one here. Aczel's stronger criterion identifies more sets than (***) and thus leaves fewer distinct sets after the identification is made. So there are more non-well-founded Finsler sets than Aczel sets. In addition, an intermediate class of non-well-founded sets was defined by Dana Scott [1960] in a lecture that was unpublished but which nevertheless existed in some libraries as a mimeographed pre-print.

In summary then, there are various possible alternatives for formulating a graph theoretic version of an isomorphism axiom of set identity. Finsler's original idea in [1926b], as amplified in (***) above, will always be of interest because it provides the greatest possible variety of sets. It gives the most liberal and expansive universe. But, as will be explained below, there are also certain advantages to accepting fewer sets as was done by Scott and Aczel.

It happens that Finsler later changed his version of Axiom II in a way that permits fewer distinct sets than he originally allowed. Thus he tended in the direction that was later adopted by Scott and Aczel. The second version of Axiom II, however, was not given in graph theoretical language at all.

Finsler's Second Version of the Axiom of Identity

The second version of Finsler's Axiom of Identity is: "The sets M and N are identical whenever possible". This means that if it is consistent to identify two sets M and N, then $M = N$. This new version of the Axiom of Identity, Axiom II, appeared abruptly in "The Infinity of the Number Line" [1954].

One needs to keep in mind that the meaning of such an axiom depends upon the theoretical context in which it appears. This becomes plain when one considers the fact that Finsler was not the first to employ this axiom.

The axiom of identity, stated just above, is equivalent to an axiom employed by Felix Bernstein in [1938] which was also entitled the "Axiom of Identity".

Bernstein added the axiom to ZFC where it tends to produce a more complete theory. For example, in a set theory with this version of the axiom of identity, the continuum hypothesis can not be independent. For, suppose the continuum hypothesis were not refutable. Then we may consistently suppose that $2^\omega = \omega_1$. Thus, using the axiom of identity, we may actually suppose that $2^\omega = \omega_1$. So the continuum hypothesis would have to be either provable or refutable.

In Bernstein's theory the term "consistency" refers to provability in Bernstein set theory itself. The theory is fundamentally circular in nature; and, in fact, this can be used to show that it is not axiomatizable as a theory in a formal language. The whole purpose of this axiom for Bernstein was that his theory should tend toward completeness.

Finsler cannot have had the same motive. His Axiom III, rather than Axiom II, is a principle of completeness. We suspect that the axiom was revised because the original version came to be seen as overly generous in the variety of sets that it allows; this is hinted at in Finsler [1964].

To judge from the experience of Jon Barwise and John Etchemendy [1987], and of Barwise [1988], in which non-well-founded sets were applied to model theoretic matters, a more restrictive universe of non-well-founded sets may turn out to be the most natural for some applications. Barwise made use of the "solution lemma" of Aczel [1988, 13]. This lemma is conveniently available for the narrow class of non-well-founded sets given by Aczel's criterion of set identity but not for the wider class of sets given by the first version of Finsler's axiom of identity.

For example, among the Finsler sets (and among the previously mentioned sets of Scott as well) there are a pair of "Fibonacci" sets that have already been introduced:

$$A = \{A, B\} \text{ and } B = \{A\}.$$

Now, in Aczel's set theory it is consistent with the membership relation to identify $J = \{J\} = A = B$. We understand Finsler's second version of Axiom II to require this identification also.

It seems a shame to lose such delightfully interesting sets. But now consider a stranger example: Let A and B be the Fibonacci sets just defined; and let $T = \{T, B\}$. The class T is a Finsler set according to the first version of the Axiom of Identity, as elaborated by (***) above. But the set equation $x = \{x, B\}$ has two solutions: A and T. Now, perhaps there is no necessary reason that sets should have to be uniquely representable as solutions to such equations involving braces; but it is certainly peculiar if they do not. And we can readily imagine that this state of affairs might make applications of non-well-founded sets distinctly clumsy.

In summary, there are at least two possible motives for changing the Axiom of Identity from its first to its second version: It allows a new kind of completeness argument; and it gives a more tractable class of sets for applications. It seems implausible to think that Finsler might have had the first of these motives. There is no reason to think that he saw Bernstein's paper which is so unknown that we do not know of any reference to it. Finally, we should like to repeat that even if one were to regard the second version as the true Axiom of Identity, the first version would retain considerable interest.

VI. *The Axiom of Completeness*

The third axiom resembles a dual of the axioms of "restriction" such as the axiom of constructibility. One might call it an axiom of expansion. It asks for the maximum class of sets compatible with the previous two axioms.

The most antithetical objection to Finsler's axioms, due to the algebraist Reinhold Baer [1928], asserted that Axiom III, the axiom of completeness, is incompatible with the first two axioms. In other words, Baer maintained that the set of all sets, A, cannot exist even as a class. Baer's argument was described by Georg Unger [1975b], but there is no English account of it. Here is the argument.

To prove: The class of all Finsler sets A does not exist. Suppose A exists. Form the Russell set

$$R = \{ x \in A : x \notin x \}.$$

Either R is a set of A or not. If it is, Russell's antinomy can be carried out in A. If it is not, then it is well-defined according to the standards of Finsler's first axiom; for the the sets of A must be well-defined, so that for any element of A either $x \in x$ or $x \notin x$. But this means that $R \in A$ by axiom III. Thus there is a contradiction. This completes the argument.

If this argument were correct then one could not form the Finsler universe A at all: Axiom III would be absurd.

The argument however does not actually apply to Finsler's theory. The definition of R does not necessarily produce a set in the Finsler theory, only a class. If R were a set, which it is not, it would have to be in the set of all sets already. How is it conceivable that a set exists outside the collection of all sets?

Finsler quickly replied to Baer's argument in [1928]. The reply was blunt: "It is perfectly clear", he wrote, "that if two deductions A and B mutually contradict one another, then it cannot be said without further ado that B is an error in deduction because it stands in contradiction to A, but it follows in the first instance only that A or B must be false."

He went straight to the flaw in Baer's reasoning: "It consists in the fact that no proof is given for the existence of the set with which the system is to be extended [...]. If one wants to "construct" a set (or any thing), one has to ask oneself whether the construction can be carried out."

In other words, Baer's argument produced a "contradiction" by adding a class to the universe while claiming that it is a set in the theory.

One curiosity concerning this objection is that it was anticipated by Finsler and is discussed in section 11 of *On the Foundations*. We can only imagine that Baer never penetrated as far as section 11 of this paper but simply believed that the axioms must be inconsistent after briefly pursuing them. It also appears that Finsler was so surprised that anyone would make this claim that he quite forgot that he had already treated Baer's line of thought.

It is odd to find that Baer adhered to his argument in a brief note [1928b] in which he claimed that R is a Finsler set simply because it is defined in terms of Finsler sets.

Finsler obviously thought that Baer's argument was of no consequence even though Baer had not admitted an error. There is a brief mention of the matter at the beginning *of The Existence of the Natural Numbers and the Continuum* [1933] which appears in this volume; he probably felt that the matter was closed. It turned out otherwise.

One of the startling things about the whole matter is that the first edition of Fraenkel and Bar-Hillel [1958, 23] endorsed Baer's argument. Their few words on the subject may have been the sole account of Finsler's theory in English. They wrote:

> From the first two axioms it follows easily that any consistent model of the Finsler system admits a further extension [...]. On the other hand, the third axiom postulates the completeness of the system [...]. But in view of the result just mentioned the third axiom entails a contradiction [...].

The "result just mentioned" refers specifically to Baer's erroneous argument. This passage, quoted above, also repeats Fraenkel's opinions of [1928b]. Now, Baer's mistaken argument can be understood as the hasty opinion of a young mathematician who avoided discussing the matter subsequently. Fraenkel's repetition of it is less easily understood.

Apparently surprised at the longevity of Baer's argument, Finsler took up the topic once more in "On the Foundations of Set Theory, Part II" [1964].

> I did not think that I would have to answer yet again in relation to the short remark of Baer [...] as surely everyone who thinks the matter over correctly must see which of the two is right. Now apparently no one has given the matter

enough thought and it has simply been concluded that he who had the last word is right.

It may be that Fraenkel sensed that there was a theorem beneath the surface of Baer's argument and, joined with a formalist bias against Finsler's Platonism, assumed that the argument was correct. If this conjecture is right, we should be able to find a correct theorem along the lines of Baer's objection.

Baer's argument can, in fact, be adapted to establish the following: If Finsler set theory is consistent, then it is not representable in the first order predicate calculus.

Only a sketch of a proof will be given here. To begin, take a model of Finsler set theory which, by way of contradiction, we suppose to be a first order theory. Let this model take the place of the entire Finsler universe, A, in Baer's argument. One needs to check that such a model for a first order theory is really a set in the sense of Finsler. Having done this, call the model M. A relativized Russell set R of Baer's argument is definable in M but not present within it. Thus the model would not satisfy Axiom III after all. Since this is a contradiction, it must be that Finsler set theory cannot be a first order theory. This completes the sketch.

One would therefore expect that someone holding an implicit belief that all consistent theories are formalizable, could jump to the conclusion that Finsler set theory is inconsistent. This seems to account for Baer's objection.

When the whole matter was brought to his attention again upon the publication of Finsler [1975], Baer declined to discuss the argument after so long an interval, nor did he agree to its being reprinted.

VII. *Consistency Proofs*

After he had stated his three axioms, in section 11 of *On the Foundations*, Finsler gave a consistency proof for them. The proof simply takes the direct union of the various classes which form partial subsystems.

In the next section it will be explained how it is that Finsler's consistency proof is so surprisingly brief. Consistency arguments are usually either reductions, in which the consistency of one theory is given relative to another, or they are proof theoretic analyses of formal systems. The very idea that a consistency argument could be both absolute and brief will seem peculiar to a formal habit of thought. So it is quite natural that a critic of this set theory should focus on the consistency argument of section 11. The rest of this section concerns a critique of this consistency argument. The whole matter is of considerable interest; and it seems to suggest that consistency proofs may ultimately contain a philosophical component.

Ernst Specker, in [1954], made a interesting criticism of the line of reasoning used in this consistency proof. Specker did not positively assert that A, the Allmenge or set of all sets, does not exist, as was attempted in Baer's mistaken argument. His objection simply casts doubt on Finsler's consistency argument. Specker's objection will be described below. The reader should first thoroughly familiarize himself with Finsler's system before considering these examples. The account given here is an abbreviated one; one cannot really judge the matter without some previous experience in the theory.

As the objection appears in print, Specker defined a class, let us call it Q_1 here, whose existence as a set is incompatible with that of A, the "Allmenge". He can then challenge the Finsler theory to produce some standards for preferring one as a set over the other. To make this challenge he presented pairs of classes which taken alone appear consistent as sets but cannot both be sets simultaneously.

Consider the following definition of a class:

$$x \in A \text{ if and only if } x = x.$$

Of course, this gives the class of all sets.

Now form the class Q_1 given by:

$$x \in Q_1 \text{ if and only if } (x \notin x) \text{ and there is a } y \text{ with } (x \in y).$$

In itself, Q_1 is consistently defined. In a partial universe it may very well be that Q_1 is compatible with all the other sets present. Let us take it to be a set then. But now, some partial universes contain A and others contain Q_1. The two sets cannot be joined in a common universe: Q_1 is a kind of dual to the empty set, it belongs to nothing, whereas A contains all. Suppose we were to have A and Q_1 present together. Then $Q_1 \in A$, so the second conjunct in the definition of Q_1 becomes true and may be dropped. We are back to Russell's paradox again.

What standards of priority are we to use in forming sets? How are we to know whether to accept A or Q_1?

Finsler mentioned this objection in [1964] and simply defended the existence of A, the set of all sets. Evidently he held that that its manifest consistency, and therefore existence, served also as a proof of the non-existence of Q_1.

Specker then modified the argument (from this point on the whole matter is entirely unpublished) by choosing a slightly different class, to be called Q_2. This revised definition no longer produces a conflict with the universal set, A, but still presents the same type of challenge to Finsler's consistency argument.

$x \in Q_2$ if and only if ($x \notin x$) and there is a y with ($y = \{x\}$).

The whole affair is essentially the same but it is no longer so easy to simply assert that all sets form singletons as it was to assert that all sets are members of something. It can be done however. Finsler, in *On the Foundations* showed that $\{x\}$ exists whenever x is a circle-free set and mentions in section 4 of Finsler [1933], translated here, that he has proven that $\{x\}$ exists whenever x exists, even when x is circular. He describes the proof as "not simple". We have not seen Finsler's argument but know another by Guerino Mazzola which is unpublished as well. Mazzola's argument constitutes a proof in the Finsler theory that Specker's class Q_2 does not exist. So the status of Q_2 as an objection to the Finsler theory is no different from that of Q_1, except that someone defending the theory has been forced to work a good deal harder.

Specker's construction in general is:

$x \in Q$ if and only if ($x \notin x$) and $P(x)$.

It can be modified by introducing other set theoretic properties $P(x)$. The resulting problem for the Finsler theory: Does Q satisfy Axiom I? may be difficult or easy, interesting or tiresome, depending on P, but all problems of this type that we have investigated have been solved. Furthermore, none of them really undermine one's confidence in the theory, because they present a pair of incompatible classes for which one can instantly tell, intuitively that one exists and the other does not. All constructions known to us that involve the formation of two equally attractive but logically incompatible classes are so artifical that the Finsler theory rejects both; the construction of J. Hintikka [1957], for example, requires a universe of bounded, finite size.

It is possible to give a very simple example which illustrates these features clearly. Consider the following class of one element:

$x \in W$ if and only if ($x = 0$) and $x = y$, for every set y.

The class W is compatible with itself alone. Were we to begin constructing the universe and happened to accept W as being the the very first creation of a set theoretic genesis, then the whole enterprise would come to a halt with no further ado. We would be left with a universe consisting of W alone.

Now, it is intuitively plain that the existence of {0} shows that *W* does not exist. Yet, if a radical nominalist were to insist that mathematics is a mere game played with marks on paper and that he sees no grounds for supposing that {0} exists in preference to *W*, is there anything in the Finsler theory that would persuade him otherwise? The attempt to answer this question has left us with the feeling that philosophical matters enter into what appears to be a purely mathematical question.

In the next section we will see that those familiar with the Finsler theory are not disturbed by the fact that it rests on a consistency argument that appears fragile to some observers. In fact, they find themselves frequently concerned with consistency in the form of showing that sets exist by the standards of Axiom I. This point of view will be elaborated in the next section. That still leaves Specker's construction as a potential source of problems for the Finsler theory and, perhaps, as a indicator of the region in which philosophical differences of opinion can arise. Finsler's theory was called "strongly impredicative" by Bernays in [1956] who added "this comment in itself doesn't signify an objection". Specker's objection is directed toward this impredicative qualities of the theory.

VIII. *Existence*

In the last section Specker's objection to Finsler's consistency proof was presented: In it two mutually incompatible classes are defined. The consistency proof itself does not contain criteria to decide between them. Finsler, in turn, regarded the decision as to which, if either, of these two objects is to be accepted as a set as something external to the argument for consistency of the axioms.

Two Swiss mathematicans, Herbert Gross and Georg Unger, an algebraist and geometer respectively, who maintained experience with Finsler set theory provided some comments on the theory upon the publication of Finsler's collected papers [1975]. From the remarks of Unger [1975a] one sees that Specker's objection to the consistency proof, is, from the perspective of the Finsler theory, actually a source of independent, set existence problems to be dealt with, but without bearing on the consistency of the theory as a whole. Unger explained the matter as follows.

> In order to follow the course of Finsler's reasoning it is indispensible to assume the possibility of speaking of all the objects (and these alone) that satisfy Axioms I and II consistently. In the wake of the foundation crisis there arose the latent question of whether individual things compatible

with I and II could be incompatible when one tries to think of them coexisting with one another, perhaps revealing hidden contradictions. I understand Finsler to mean that one does not accept only those things that satisfy Axioms I and II alone but those which are also free of contradiction among themselves. The question of whether or not such a system is accessible to the human mind is answered in a concrete way through presenting the empty set along with other elementary structures as examples. Certainly they are in the union of all mutually compatible objects satisfying Axioms I and II. Thus it is non-empty.

IX. Circular Sets

Any set, S, that contains itself is circular. Even if some set essential in S contains itself, we should regard S as circular, because it is ultimately dependent on a membership cycle. The problem now is to say which of the remaining sets, those of the class Σ^*, to use Finsler's notation, are also circular. Their circularity depends on the definitions which characterize them. A definition must be something that can actually be held in thought, otherwise it would not give a set according to Axiom I. But we do not mean to specify a particular syntactical form that a definition must take: a definition is "conceptual" rather than "linguistic" to use terms from the philosophical section of this book.

Of course, a set could have many definitions, of quite varied kinds. The common, familiar sets possess linguistically expressible definitions, a fact which stands to the commercial benefit of the manufacturers of pencils and sticks of chalk.

Returning to the concept of a circle-free set, we have seen that certain sets, those outside of Σ^*, are obviously circular. In section 13, Finsler defines circularity for the sets of Σ^* too. These are said to be circle-free if everything in their transitive hulls is independent of the concept "circle-free"; otherwise they are circular.

Some readers find Finsler's definition thoroughly confusing. It may help to substitute some other expression for the term "circle-free" as it appears in quotation marks. For example, let us introduce a new, ideal object into set theory and call it "X". It helps to think of it as predicate, though the word "predicate" has an unintended syntactical connotation. In any case, set theory now involves membership (the denotation of $\in$) and the mysterious X class, which we may indicate by writing "X". Certain classes, the class X itself for example, are definable using X. Others, like the empty set, are definable without reference to X. Of course, we have at the back of our minds that X is the circle-free sets. But we do not need to say or even think so.

Any class can now be used as an expansion of set theory. That is, let C be a class and A the class of all sets. Any definition, $Def(X)$, possibly involving X can be carried out with X understood as being equal to C. Let us say that our definition gives $Set(C)$ when $X = C$.

To repeat, $Def(X)$ is a definition involving the unspecified notion X. When X is understood to be C, then this definition may yield a specific set, provided that Axioms I and II are satisfied; in this event let $Set(C)$ be that set.

If $Set(C)$ is the same set regardless of what C may be, then we shall call $Set(C)$ "circle-free", because it has a circle-free definition, one that does not really depend on X. It may also have circular definitions, those that vary as X varies, but possessing circle-free definition makes a set circle-free.

The usual definition of the empty set shows that it is circle-free. This definition is even expressible in first order set theory with bounded quantifiers and without any occurrence of "X" taken as a predicate in an expansion of the language of first order set theory. Such classes are invariably circle-free sets.

Let U be the class of circle-free sets. One can proceed to show that it is a set by Axiom I. It certainly is circular; as a set it may enter into the definition of other sets. Whenever these newly defined sets really depend on U (perhaps the class X is involved in their definitions) then they too will be, in general, circular.

The empty set also has circular definitions. Let T be the set 0, if $0 \notin X$; and let $T = \{0\}$, if $0 \in X$. This definition, call it $Def(X)$ defines different things, depending on X. If $X = 0$, then $T = 0$. If $X = A$, the "Allmenge" or universe, then $T = \{0\}$. So $Def(X)$ simply does not give us a circle-free object. When we put $X = U$, however, we have $0 \in X$ by our previous observation that 0 possesses a circle-free definition. Thus we may say that $T = 0$, when X is the particular class that it was intended to be. This whole affair is merely a circular definition of the empty set.

We believe these ideas are the most important part of this foundational portion of Finsler's work. They seem to enter into unexplored territory, except, perhaps, for the treatment of Ackermann.

X. Ackermann's Theory

The set theory of Wilhelm Ackermann [1956] was an attempt to find axioms which are closer in spirit to the ideas of Cantor than the usual set theories. Ackermann's theory, however, unlike Finsler's is formalized and attempts to avoid unsatisfiable circles by syntactical devices. There are other differences too.

In Ackermann's system, for example, a subclass of a set must be a set. While in Finsler's theory the mere fact that something is included in a set does not in the least guarantee that it is itself sharply defined. The Allmenge, A, is a set and certainly many of its subclasses, the class of the Russell paradox conspicuously among them, do not form bona fide sets.

Once we leave the realm of well-founded sets the Ackermann theory offers no guidance. There is, for example, no way to prove even that the J-set, $J = \{J\}$, exists, either as a set or a class. Probably for this reason research into the Ackermann theory, W. Reinhart [1970] or C. Alkor [1982] for example, tends to treat the theory with an axiom of foundation, either for sets or classes, adjoined.

These differences notwithstanding, the Ackermann theory often reminds one of Finsler's. To see that there is a relation between the two, take Ackermann's predicate "M", a primitive concept holding for sets, to refer to the circle-free sets of Finsler. That is, look at Ackermann's theory as a formalized approximation to the study of circle-free sets, U. Now, the parallels between the two become more striking. The class of all sets in Ackermann is not a set: The set of circle-free sets in Finsler is not circle-free. The axiom of regularity was not used by Ackermann: The axiom of regularity is false in Finsler set theory. The axiom of infinity was derived by Ackermann using an argument identical to that of Finsler. A peculiarity involving the axiom of replacement in Ackermann's theory noticed by Azriel Levy [1959] manifests itself in the circle-free sets too.

This relation has been noticed before. Paul Bernays in [1956, 264], the same year in which Ackermann's paper appeared, wrote:

> [...] the mentioned separating off of the circle-free sets can also be found in an analogous procedure – made clear through formalization – in a recent axiomatization of set theory by Ackermann.

Finsler probably thought that Ackermann had specifically attempted to treat the circle-free sets without saying so. In the same discussion from which Bernays was just quoted Finsler [1956b] replied:

   That the concept of circle-free sets, indispensible for a full understanding of mathematics, has found a certain amount of attention after thirty years is also very pleasing. What is called "made clear through formalization" points to Ackermann's deduction of a formula corresponding to the axiom of infinity, using specific formulas which refer to the property of being a circle-free set. The actual existence of infinitely many things cannot be guaranteed in this way. This turns mathematics into "a doing as if", pretending that there are really infinitely many things. I cannot accept this.

Of the various studies of the Ackermann theory it is the investigation of natural models as in Rudolf Grewe [1969] and John Lake [1975] that are most related to issues that arise in Finsler set theory. Even though these natural models do not take place in a context in which the Axiom of Completeness, Axiom III, holds, they do at least embrace an initial segment of the Zermelo-Fraenkel universe.
   Herbert Gross [1975, vii] in the preface to Finsler [1975] wrote: "An axiomatic set theory which refers to objective mathematical things, the conceptually fascinating idea of circle-free sets, and Finsler's approach to the foundation of mathematics are all worthy of investigation." We have also found that the history of these ideas is fascinating as well.

XI. Large Cardinals

   One of the mathematical observations in Part II [1964, §61] is that there are circular ordinals. Among the consequences is the absolute existence of an inaccessible cardinal.

First published as: "Über die Grundlegung der Mengenlehre, Erster Teil. Die Mengen und ihre Axiome", *Mathematische Zeitschrift* 25 (1926), 683–713.

On the Foundations of Set Theory

Part I. Sets and their Axioms

Introduction

Pure mathematics operates with objects which are either immediately given, determined axiomatically, or with objects derived from these by specific constructions.

Examples of such objects are: the natural numbers; the real or complex numbers; the real or complex functions, in particular the analytic functions; the points, lines and planes of Euclidean or projective spaces; and so on.

Each of these examples, understood in the usual way[1], concerns well-defined systems, which are in no way merely arbitrary.

A system of similar exactitude has not yet been found for set theory.

On the one hand, one must very often admit quite diverse elements into our sets, without ever limiting the domain from which they are chosen: On the other hand, one seems to obtain such a great variety of sets that they eventually cannot be collected together, even when starting from a fixed domain.

Restricting the formation of sets by axioms such as those given by Zermelo [1908] offers no guarantee of a fixed universe of sets. Even if one assumes that such a universal system exists, it would certainly be dependent upon the arbitrary choice of the axioms. These disadvantages do not disappear if, following Fraenkel [1922a], one adds an "axiom of restriction" (cf. Skolem [1922], von Neumann [1925]).

The "antinomies" which arise from the construction of "the set of all sets" and similar conceptual formations constitute a fundamental obstacle for the foundation of a general set theory which is free of contradictions. Without a clarification of these one can hardly conceive of a satisfactory foundation for set theory.

---

[1] In this work we adopt the standpoint that exact mathematics includes the law of the excluded middle. The investigations of L.E.J. Brouwer and H. Weyl, which move in another direction, are therefore outside of our scope.

I have, therefore, attempted to undertake such a clarification in [1925] and should like to add to this a few supplementary remarks. It will then become apparent that a universal system of sets can be defined that fulfills the requirements. Sets will be introduced purely axiomatically as ideal objects between which a certain relation holds, rather than explicitly defined. Further, by means of this system, a distinction is achieved between "circular" sets, from which paradoxes can arise, and "circle-free" sets, which are important for set theory proper and for the rest of mathematics. This then also furnishes the means with which to investigate the validity of the axioms of Zermelo and similar principles.

The consistency of arithmetic and analysis, and the foundation of transfinite ordinal numbers, will be elaborated from these principles in the second part of this investigation.[2]

The final aim being pursued in this matter is exactly the same as that which Hilbert has stated in his work [1922] and [1923] towards the new foundation of mathematics: Here, however, the question of consistency will be understood in an absolute sense, not a formal sense as in Hilbert. In order to found an exact science properly one must acknowledge an absolute logic upon which to support it ´and without which no rigour in proofs is conceivable. In particular the law of the excluded middle will be included in this logic; for example, a real number is either rational or irrational even when a decision procedure for this can never be accomplished by human means. Thus a statement can also be true, even if it cannot be proved so in a "finite number of logical steps"; and similarly a system can contain a contradiction, even though one is certain that a finite method of proof would never reveal it (see Finsler [1926a]). Besides, the concept of "finite number" must not be taken for granted, especially not in a theory which is supposed to provide a basis for it.

Even though the approach developed here is different from that of Hilbert, I believe nonetheless that it does lie quite legitimately within the domain of the "axiomatic method". At this stage I should not like to omit to express my most cordial thanks to my teacher David Hilbert for the rich stimulation which I experienced, particularly from his lectures directly concerning these matters.

I am especially obliged to Paul Bernays for critical remarks and valuable advice in connection with the development of this work.

---

[2] *Editor's Note*: Controversies surrounding Finsler's axioms altered his plans for the second part of this paper. However, see Finsler [1933], [1941b], [1951], [1954], [1964].

Chapter 1. The Antinomies

§1. A False Assumption

The assumption that arbitrarily specified things can always be combined together into a set, in other words that there always exists a definite set which contains specified objects and only these as its elements, underlies the naive set theory which leads to the antinomies. By this assumption sets are among the things which can occur as elements of sets.

Let the relation of a set to its elements (i. e. the relation of "containment") be denoted[3] by β.

The assumption of naive set theory can now be stated: For arbitrarily given objects there always exists a uniquely determined thing which possesses the relation β to the given objects and only to these.

This, however, is a postulate which necessarily leads to contradictions; for there is no domain of things in which it can be satisfied without restriction.

That is to say, in any domain of things, in which the relation β is given, one need only take those things A of the domain which do not possess the relation β to themselves; that is, those for which $A\beta A$ does not hold. Then there is no object N in the domain possessing the relation β to exactly these things, because both the assumption $N\beta N$ and its negation lead immediately to a contradiction.

Applying this postulate to a domain containing one element, we obtain a single thing J, which possesses the relation β to itself.

In a domain with more than one element, however, the naive postulate cannot be satisfied. That is to say, if a and b are any two different things of the domain then there must exist in the domain three distinct things A, B, and C such that A possesses the relation β to a only, B only to b, and C to both a and b. Then, however, $A \beta A$, $B \beta B$, $C \beta C$ cannot all hold simultaneously, for, from this it would have to follow that $A = a$, $B = b$, $C = a$ or $C = b$, whereas it is most certainly the case that C is different from A and different from B. Hence there would certainly exist things which do not possess the relation β to themselves, but then once again the domain could not contain a thing N that possesses the relation β to exactly these things.

---

[3] The relation ß is converse to the relation ∈ of Zermelo [1908], i.e., $M \beta x$, just as $x \in M$, means that x is an element of the set M.

§2. *Circular Definitions*

It follows from these considerations that the naive assumption cannot be valid.

In spite of this, the fact that one is accustomed to comprehending a totality of many things as a unity stands in favor of the assumption. A collection is often thought of as something given in itself.

This assumption is possible only as long as *circle-free* operations alone are employed, as in those cases in which the things to be collected together are not themselves collections or dependent on collections.

The case of *circular* collections is different, however. The set of all sets, for example, must be a collection which contains itself. Under certain circumstances such circular constructions can be uniquely satisfiable but there may also exist other conditions under which the construction becomes indeterminate, and in these the circle is not satisfiable. All of these possibilities must be taken into consideration.

The situation here is similar to that in algebra where an equation of the form $x = f(x)$ possesses a unique solution under certain circumstances, while under others the number x remains indeterminate, and finally it may even be that the equation is not satisfied by any value of x at all and is therefore insoluble. If for example, the equation

$$x = a + bx$$

is taken, then each of these cases occurs, according to the a and b that are chosen: $a = 1$, $b = -1$; or $a = 0$, $b = 1$; or $a = 1$, $b = 1$. For $a = 1$, $b = 0$ the definition becomes circle-free.

Since the naive definition of the concept of a set is of a circular in nature, all the above cases can occur also in connection with it. The assumption that any things whatever can invariably be collected together is no longer tenable. If one were to collect together exactly those collections which do not contain themselves, then one would require something impossible; this is so *precisely* because the circle involved is not satisfiable.

From this it also follows that a set is not necessarily given whenever only its elements are given. Rather, the set must first be formed, and it can only be formed with complete assurance if one knows that no circle arises.

It has often been called to special attention that a set which contains a single element, or, as one says, which "consists of one element", must not be confused with the element itself. Still less is it permissible to confuse a set which contains many elements with the totality of all these elements. "Many things" are not one thing and cannot be identified with a single thing. Generally, however, one can associate with them a single thing. One can always do this as long as unsatisfiable circles do not arise.

Just as in ordinary life it is always the circle-free cases which occur, so here too a composite thing will usually not be distinguished from the totality of all its separate elements even though this is not the same thing. A forest conceived as a unity is not identical with the trees from which it is "composed": We can indeed go *into* the forest, but only *between* the trees!

The distinction is of importance for set theory, because on one hand the number of elements enters in, and on the other hand, it can happen that the elements do indeed exist individually while the corresponding set cannot be formed because of a non-satisfiable circle.

§3. Sets and Classes

It is good to bring this distinction into our vocabulary. It would surely be inconvenient if one always had to speak of many things in the plural; it is much more convenient to use the singular and speak of them as a *class*. This usage will of course be adhered to in what follows and will not lead into difficulties, provided that one pays attention to the fact that "class" does not have the same meaning as "set". A class of things is understood as being the things themselves, while the set which contains them as its elements is a single thing, in general distinct from the things comprising it.

For example, if the proposition which states that any number possesses a unique decomposition into prime factors is valid for the class of all natural numbers, then it is valid for every single natural number. If, however, the set of all natural numbers is an element of a set M, then it could be that M possesses only this single element; the individual numbers would not be elements of M.

Thus a set is a genuine, individual entity. By contrast, a class is singular only by virtue of linguistic usage; in actuality, it almost always signifies a plurality. In particular, the word "class" can also be used in the special case that the number of things collected is equal to one or even zero. "Totality", as occurring in the foregoing discussion, has been employed with similar meaning to that of the word "class" adopted now.

With this the origin of the set theoretic antinomies can now be explained by saying that to every class of things there does not necessarily correspond a set. Thus one can truly speak without inconsistency of all sets which do not contain themselves as elements, *and also* even of the class of all such sets; but there does not exist a corresponding set.

One could argue that in speaking of a "class" of things, or even by saying "all" things endowed with a certain property, then it is already the collection itself which is meant and that such a collection should be called a set. But to speak of all sets which do not contain themselves, is really contradictory. The contradiction originates from the fact that this mode of expression rests on an impossible concept.

This latter standpoint, however, is unnecessary. The situation becomes much more clear when the concept of a set is modified, so that even a well-defined collection of things does not necessarily form a set. On the contrary, sets are things which *correspond* to collections, in so far as this is consistent. It is in general better not to refer to all collections as "things".

Adopting these conventions, one can speak of a class of sets without having to connect the formation of a set with it; otherwise one needs to be certain that there is no unsatisfiable circle.

Various concepts which are defined for sets also are meaningful for classes. For example one can assert of the class of all sets which do not contain themselves that it is uncountable. One can ask whether it is well-ordered. Counting and ordering only require the "many things", the elements of a class "in their entirety", and not the one thing, the set, which in this example certainly does not exist.

§4. Pure Sets

As soon as one has convinced oneself that a collection of arbitrarily specified things need not always correspond to a set (this is the case whenever the formation involves a non-satisfiable circle), then the well known paradoxes disappear.

There now arises the question as to whether a complete foundation for set theory can be obtained on these grounds.

The concept of a collection of arbitrary things is vague as long as a "thing" has not been precisely defined. In order to achieve a sharp definition one can proceed so as to exclude everything which is not absolutely necessary, i. e., everything except the sets themselves. One thus obtains "pure sets", whose elements are themselves always again only pure sets.[4]

---

[4]   In 1920 I communicated to P. Bernays the idea of adopting the system of these pure sets as a basis for the investigation. The axioms of Fraenkel ([1923] and [1925]) likewise

In order to completely avoid the previous difficulties, which lie in the concept of collection itself, and so to obtain firm fundamentals, we will determine these sets purely axiomatically.

In mathematics a *point* is not "that which has no part" but is an ideal element which satisfies certain axioms though it is connected by various analogies to what one thinks of intuitively as a point.

A *natural number* is, in my opinion, not a mere symbol but is an ideal element, which satisfies certain axioms, and to which the symbols used in calculation possess a close analogy.

So too, a *set* should not be only a collection but an ideal thing which satisfies certain axioms, and which stands in close analogy to the intuitive collections of naive set theory, and in particular with the sets which were called "pure sets" above.

The existence of the things in this system is based on their consistency. The system can then be used as a basis for set theory and for the rest of mathematics.

From this axiomatic point of view sets are things in themselves rather than collections; it will soon be apparent that we can collect sets together without risking the danger of circular constructions. The collections themselves, however, cannot then be referred to as sets but only *systems* of sets. By a *class* of sets we still mean the sets which are collected. The distinction between systems and classes is, however, of no great importance once the concept of a set has been grasped.

Since the systems of sets are now no longer fundamental objects which could then enter into sets as elements, it is consequently pointless to form systems of systems and so on, in an arbitrary fashion, and thereby become entangled afresh in new circles.

Chapter 2. The Axioms of Set Theory

§5. The Axiom System

After these preliminary observations we can say in keeping with Hilbert's [1913; 2, 238] axiomatics (as will soon be apparent, the three axioms are *free assumptions*, no distinction is made here between "axioms" and "postulates"):

---

use only such sets, but they seek to exclude the antinomies by means of additional restrictions, as do those of Zermelo.

We consider a system of things, which we call sets, and a relation, which we symbolize by β. The exact and complete description is achieved by means of the following axioms:

I. Axiom of Relation:
For arbitrary sets M and N it is always uniquely determined whether M possesses the relation β to N, or not.

II. Axiom of Identity:
Isomorphic sets are identical.

III. Axiom of Completeness:
The sets form a system of things which, by strict adherence to the axioms I and II, is no longer capable of extension.
That is, it is not possible to adjoin further things in such a way that the axioms I and II are also satisfied.

The expression "isomorphic" will be defined in §7. Let the system which satisfies these axioms be denoted by Σ; then the word "set" only means "a thing of the system Σ". The sets themselves must be purely ideal things, which are determined solely by means of the relations which are stipulated between them.

In the following $M, N, A, a, \ldots$, will denote sets; $\Sigma, \Gamma, \ldots$, systems of sets; equality between sets or systems ($A = B$, $\Sigma_1 = \Sigma_2, \ldots$) signifies identity.

§6. The First Axiom

Consider, in the first instance, an arbitrary, but fixed system Σ for which the first axiom is satisfied. The word "set" shall mean, of course, no more than "a thing belonging to the basic system Σ under onsideration".

If a set M possesses the relation β to another set A or in short $M \beta A$ holds, then A shall be called an *element* of M. We shall, in general, retain the designations which are customary in set theory and therefore say, for $M \beta A$: The set M "contains" the set A. The set M can at the same time contain still further elements $B, C, D, \ldots$ but it follows from axiom I that the class of all these elements is always uniquely determined.

If M is given, then one can also say conversely, that the sets $A, B, C, D, \ldots$ together "form" the set M, though one must not allow oneself to be led astray into false ideas by this expression. Arbitrary sets taken together need not always form a set, i. e., it is nor postulated that there always exists a set which has the relation β to some given class.

For certain purposes it is generally advantageous to avoid introducing the intuitive notion of "containment" into the symbol β, which should represent solely a relationship between things and which one can also think of as being represented by an arrow (examples are given below). The situation here is roughly similar to that relationship which the number $n + 1$ bears to the number n.

Thus there is no difficulty involved even in introducing a set which "contains itself as an element", and in particular a set which contains *only* itself. This is simply a thing of the system Σ which possesses the relation β to itself, and to itself alone. That something can possess a relation to itself is nothing unusual: Similarity is a relation which a geometrical figure possesses to itself too; identity is one which something possesses only to itself.

A thing of Σ which possesses the relation β to nothing at all (not just to no other thing of the system) shall be called an *empty set*. It is a set, containing no elements, which therefore corresponds to the usual empty set. In the present theory however it possesses, if it exists, exactly the same status as any other thing of Σ. By contrast the situation in the usually formulated set theory is that the empty set possesses a singular status, being "a collection which collects together nothing at all".

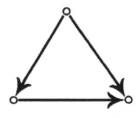

Figure 2

Anything of Σ which possesses the relation β to an empty set alone, which thus contains an empty set as its sole element, will be termed a *unit set*.

Let the relation β be represented by arrows and the sets by circles from which these arrows proceed. An empty set would then be represented by a circle having no arrow. A unit set is given by a node having only one arrow directed toward an empty set. The upper circle in Figure 2 represents a set which contains an empty set and a unit set.

Figure 3

A set which contains only itself is represented by a circle from which an arrow proceeds looping back to the circle (Figure 3): the J-set.

If one uses bracket notation then the empty set and the unit set can be represented as: { } and {{ }} respectively, whereas for a set which contains itself equations, $H = \{A, ..., H\}$ or $J = \{J\}$, are necessary.

§7. Transitive systems

Let Σ be a system of sets satisfying axiom I. For example one could take an empty set, a unit set, and a J-set.

Arbitrary things of Σ can be collected together into a *subsystem*. Any such subsystem is then well-defined whenever, for each set, it is uniquely decided whether it belongs to the subsystem or not.

A system of sets which contains all the elements of its members will be called *transitive*. The system Σ in which axiom I holds shall be taken as transitive.

Consider an arbitrary set M of the system Σ. There is at least one transitive system that contains all elements of M, namely Σ itself. Those sets which belong to *every* system of this kind are said to be *essential in M*.

The system of sets which are essential in M will be denoted by Σ_M. It is uniquely defined; for if N is an arbitrary set then either there exists a transitive system which contains all the elements of M but not the set N, or there does not exist such a system. In the second case, and only in this case, does the set N belong to the system Σ_M.

As a consequence of the definition, every element of M is essential in M. Further, let the set A be essential in M and let a be an element of A, then a is also essential in M; for there is no transitive system which contains the set A but not the set a. From this it follows that Σ_M is also a transitive system that contains the elements of M. We have therefore.

Proposition 1. *The sets which are essential in some set M form a transitive system which contains all the elements of M. The same is also true when the set M is added to Σ_M.*

Sets which are essential in some set M belong, by definiton, to every transitive system which contains all the elements of M; and so in particular they also belong to every transitive system which contains the set M. This gives:

Proposition 2. *A transitive system contains all those sets which are essential in any of its elements M.*

From this we have, as a corollary:

Proposition 3. *If A is essential in B, and B is essential in C, then A is essential in C.*

By proposition 1 the system Σ_C of sets essential in C is a transitive system. By hypothesis it contains the set B and so consequently, by proposition 2, also the set A; this completes the proof.

The system of sets which are either identical to an element of M, or which are essential in an element of M is a transitive system which contains all the elements of M and therefore also all sets which are essential in M. Thus we have:

Proposition 4. *If A is essential in M, then A is either an element of M, or is essential in some element of M.*

Since conversely, by proposition 3, every set which is essential in some element of M is also essential in M itself, the system of sets essential in M is identical to the system which consists of the elements of M and those sets which are essential in these elements.

Any one-to-one mapping which maps two transitive systems onto one another is called *relation preserving*, if whenever the relation $A \beta a$ holds for sets of the one system, then the relation $A' \beta a'$ always holds also for the associated sets of the other system and conversely.

Two sets M and M' are *isomorphic* if the system $\Sigma_{\{M\}}$ and $\Sigma_{\{M'\}}$ of sets which are essential in $\{M\}$ and $\{M'\}$ respectively can be mapped onto one another by a one-to-one relation preserving mapping in such a way that the elements of M are mapped onto those of M'. That this latter condition is not superfluous is shown by the example

$$A = \{A, B\}, \quad B = \{A\},$$

where A is not isomorphic to B.

Every set is isomorphic to itself and two sets which are isomorphic to a third are also isomorphic to each other.

§8. The Second Axiom

From now on suppose that an initial system Σ satisfies the second axiom, which states that isomorphic sets are identical. In other words, for each set in Σ there shall not exist any other which is isomorphic to it.

If the sets M and M' possess the same elements, then according to proposition 4, the systems Σ_M and $\Sigma_{M'}$ are identical. One can correspond M and M' with each other, so there is an isomorphism between $\Sigma_{\{M\}}$ and $\Sigma_{\{M'\}}$. This means that M and M' are isomorphic so by axiom II we have:

Proposition 5. *Two sets which possess the same elements are identical.*

It is not sufficient, however, to replace the second axiom by this statement. This can be seen by means of the following example (cf. also §18).

The system Σ contains a set J, which contains itself as its sole element, $J = \{J\}$.

Suppose that K is any other set which likewise contains only itself and which consequently satisfies the equation $K = \{K\}$.

Now suppose that $J \neq K$. Then the elements of the two sets are different, and according to proposition 5, so are the sets themselves as well. The system could now contain even more sets of this sort, which would likewise possess only themselves as their sole elements and which would, with equal justice, have to be considered different. Indeed, these sets could now arise in quite unlimited numbers, higher than any of the usual cardinal numbers[5]; this would serve no useful purpose whatever. Moreover, since the assumption $J = K$ would not contradict the axioms, the question as to whether J is identical to K would not be decidable.

Now by axion II, however, all these sets must be identical, as they are isomorphic to one another. There can exist at most *one* set which contains itself as sole element. This set will be denoted as the *J-set* throughout the following.

From axiom II, or from proposition 5, it follows further that there can exist only *one* empty set and likewise only *one* unit set. These are, however, different from one another and from the J-set, since they are not isomorphic, as can easily be seen.

---

[5] Not, however, higher than the cardinal number of the "set of all things", in so far as this non-axiomatic set is admitted.

If the sets A, B, C satisfy the relations: $A = \{B\}$, $B = \{C\}$, $C = \{A\}$, then they are isomorphic, and hence by axiom II are identical to one another, and consequently to the J-set.

In general, therefore, one can conclude that if the basic system Σ satisfies the first axiom, then it is unambiguously decidable whether it also satisfies the second axiom or not. If it does also satisfy the second axiom, then it is once more unambiguous whether arbitrarily given sets are isomorphic or not, and consequently whether they are identical or not.

§9. Union and Intersection of Systems

Suppose now that arbitrarily many initial transitive systems of sets Σ_α are given, each satisfying axioms I and II. A system Σ is then said to be the *union* of these systems if it satisfies axioms I and II and in addition possesses the property that every set in Σ is isomorphic to at least one of the given sets and conversely each of the given sets is isomorphic to some definite set of the system Σ. Here the expression "isomorphic" is defined just as it was previously and it is unterstood that sets isomorphic to a given set are isomorphic to each other.

One can now obtain a union system Σ by joining together all the things which occur in the systems Σ_α and adding the condition that all the sets which are isomorphic to one another (and only such sets) are to be considered equal. Accordingly, a set M in Σ possesses the relation β to some other set N if and only if this is the case in one of the given systems Σ_α.

The stipulation made here is admissible. There is nothing contradicting it through which some of these isomorphic sets might not be identical. For a single system Σ_α, it is valid by hypothesis; even if some other general principle were taken for isomorphism between sets from different systems, then it could be ignored without affecting any Σ_α. If, therefore, for example, an empty set occurs in several of the given systems, then it is to be considered as being "the" empty set, which belongs to all of these systems.

One finds accordingly that to any collection of systems Σ_a there always exists a union Σ. It is essential to realize that in forming this union Σ the existence of things which did not already exist previously is not postulated. The system Σ could even be identical with one of the initial systems.

There is only *one* union system, Σ; for if Σ' were any other such system, then every set in Σ would have to be isomorphic to some definite set of Σ' and conversely, i. e., the systems Σ and Σ' could be mapped onto one another by a well-defined, one-to-one relation preserving mapping.

Just as a system Σ, for which the axioms I and II are satisfied admits a union, so also it admits intersections. Unions and intersections of transitive systems are transitive. By propositions 1 and 2 the intersection of all transitive systems which contain a given set M, consists of the set M together with all sets essential in M.

§10. The Third Axiom

Axiom III is the counterpart of Hilbert's axiom of completeness [1913; 22, 240]. For our purpose it says that there is a largest possible system Σ, in which axioms I and II are both satisfied.

A consequence of this axiom is:

Proposition 6. *For any well-defined class of sets, there exists a set which contains each member of the class, if and only if the assumption that such a set exists does not contradict axiom I.*

The "if" direction alone requires proof. If this assumption does not contradict axiom I, then there must exist a transitive system which satisfies the first axiom and which contains a set M possessing the relation β to just the given sets. The sets essential in M would then form a subsystem Σ_M of Σ. Adjoin M to Σ_M obtaining Σ'. This system, Σ', is also a transitive system satisfying axioms I and II, so the set M, or one isomorphic to it, must already be in Σ. Otherwise Σ could be extended by joining Σ' to it.

The third axiom can, however, not be replaced by proposition 6; for since the word "set" refers only to things of the system Σ, it would not be necessary for the J set to belong to Σ. The system Σ would then not be well-defined.

The following statement, however, is equivalent to axiom III:

Proposition 7. *An arbitrarily defined set M exists (i. e., there exists in Σ a set M satisfying such a definition) if the assumption that such a set M exists does not contradict the first two axioms.*

Using axiom III there exists a system which satisfies the first two axioms, and which contains the set M. The union of this system with Σ would constitute an extension of Σ, if Σ did not already contain the set M. Conversely, axiom III is a consequence of proposition 7; for if it were possible to extend the system Σ in accordance with proposition 7, then every new set so obtained would certainly belong to Σ.

From axiom III it follows, in particular, that the system Σ is non-empty; for the empty set, the unit set, and the J-set must certainly all belong to Σ. Further sets can easily be constructed. The assumption of the existence of a "set of all sets" does not contradict the first two axioms, as from these alone it cannot follow that a set needs to be excluded from being an element of such a set of all sets. So by proposition 7 the system Σ must also contain one such set (cf. §12).

It must now, however, still be shown that the system Σ, which satisfies all these conditions, really does exist; in other words, that the axiom system itself is consistent.

§11.   The Consistency, Completeness, and Independence of the Axioms

The axiom system set up in §5 is consistent, i. e., from the given axioms no contradiction can be deduced logically. In order to see this, let us construct a system of things in which the three axioms are satisfied.

Start with arbitrarily given systems Σ_α in which the axioms I and II are satisfied. That such systems do exist is shown by the example of a system consisting of the empty and unit sets, i. e. a system which consists of two different things, of which one possesses the relation β only to the other, whereas the other possesses the relation β to nothing at all.

In §9 it was shown how arbitrary systems Σ_α can be united to form a single system Σ, in which the axioms I and II are also satisfied. We now form, in this fashion, a system Σ which is the union of *all* possible systems Σ_α in which the axioms I and II hold. To speak of all these systems entails no circle, as it is only a system of sets, and not a system of systems with which one is dealing. The union can be obtained in exactly the same way as was given in §9.

The system Σ so formed now satisfies also the third axiom, since every system Σ' which satisfies axioms I and II must occur among the systems Σ_α. Therefore Σ' cannot contain any set which is not isomorphic to some set of the system Σ, that is, Σ' cannot be a proper extension of Σ.

Further, it follows from these considerations that the system Σ is *uniquely* determined by means of the three axioms, i. e., every system Σ' which satisfies all three axioms can be mapped onto Σ by a one-to-one relation preserving mapping. Were this not the case then either Σ or Σ' would contain a set which was not isomorphic to any set of the other system; the union of Σ with Σ', constructed as in §9, would then constitute an extension of at least one of these systems and therefore axiom III would not be satisfied by this system.

The consistency and completeness of the axiom system are secured by these means. Every unambiguous question relative to the system possesses an unambiguous answer, irrespective of whether this is obtainable by human means or not. Should two essentially different answers exist then these would constitute two contradictory statements and thereby reveal an inconsistency of the system Σ, which is not possible.

The independence of the axioms results from the following examples:

The first axiom is not satisfied by a system consisting of two things each of which arbitrarily possesses the relation β either to the other or to nothing at all.

A system which consists of two non-identical things each of which possesses the relation β to nothing at all, satisfies the first but not the second axiom.

A system consisting of the empty and unit sets satisfies both axioms I and II but not axiom III.

The axioms are, therefore, independent in the sense that none can be deduced from the preceeding; the second axiom has, however, no well-defined meaning without the first, and likewise neither has the third without both of the first two.

§12. Objections

With the antinomies in mind a few objections which could be advanced against the system Σ will now be discussed.

According to Cantor a set always possesses a smaller cardinal number than the set of all its subsets. This, however, cannot hold for the set of all sets, since it has to contain all its subsets as elements. Yet we have come to the conclusion that such a set of all sets does exist in the system Σ.

Indeed it is here that the proof of Cantor breaks down; for he supposes that an arbitrary class of sets always forms a set, and such an hypothesis is not satisfied in Σ. On the contrary, one can conclude just the reverse. By applying the Cantor diagonal argument to a correspondence between the set of all sets and the set of all its subsets, with which it is identical, the diagonal class one obtains is never iself a set. The simplest diagonal class is formed from the sets which do not contain themselves. The assumption that there exists a set which contains these and only these things as elements contradicts axiom I. According to this axiom it would either have to possess the relation β to itself or not, neither of which is possible.

In contrast to this it is unambiguously decided that the sets of all sets possesses the relation β to itself and to every other set. For every other set too it must be definite whether it possesses the β relation to any one of the sets, otherwise it is not allowed to occur in Σ. The class of all sets on the other hand is uniquely defined, so the existence of the set of all sets can be taken from proposition 6. In the system Σ this set does possess the greatest cardinal number.

One can, however, sharpen the objection to the system Σ in the following way:

Suppose that the system Σ satisfies axioms I, II and III. Now consider the class of things of Σ which do not possess the relation β to themselves. We have already observed that there does not exist in Σ any set which possesses the relation β to all and only these things. Now take a new thing N that does not belong to Σ and stipulate that it is to possess the relation b to the aggregate just considered. If this thing N is now added to the things of Σ, then one obtains a new system Σ' which represents an extension of the system Σ, in contradiction to axiom III.

In this objection there is an error, however (see Finsler [1925]). The definition of N which is given is not correct; it contains an inherent contradiction. Namely, if N is something "new", i. e., something which does not belong to the system Σ, them it is not permissible for it to belong to any system satisfying both of the axioms I and II. The condition contradicts itself; there cannot exist any such thing N satisfying it, and so, for that reason, an extension of the system Σ is also not possible.

Now and then the question is raised as to why certain axioms should be admitted in set theory, while the axiom that arbitrary things form a set should be rejected. That this axiom necessarily leads to contradictions was shown already in §1. In our system, it would assume the following form:

I*. *For every fixed class of sets there always exists a set, which possesses the relation β to the elements of this class.*

Now the essential difference between this and the axioms set up in §5 is that in I* an existence is postulated – the existence of certain things, which are to satisfy the given condition – without considering whether this condition can be satisfied. But there arise cases, in which I* leads into a non-satisfiable circle and then it requires something impossible.

The axioms I, II, and III, however, do not require the existence of things which do not already exist consistently. Only *if* things exist having certain properties can they be admitted to the system Σ. As a consequence Σ exists and is consistent.

The objections which Skolem [1922] has raised against the axiomatic foundation of set theory cannot be directed against the present system. In particular it has to be understood that in this system all concepts must be taken in an absolute sense. If it can be shown that the system Σ is uncountable (cf. §18) then this property is absolute and it cannot be realized in any countable domain by making special stipulations. Admittedly an absolute logic has to be acknowledged and with that the existence of things which cannot be individually defined in a finite way must also be permitted. Mathematical objects and their properties must be independent of any limitations which might be imposed upon them by the possible modes of representation.

Chapter 3. The Formation of Sets

§13. Circle-free and Circular Sets

Since one cannot consider every class of sets as providing a new set, one needs other rules which allow the construction of particular sets, principles by means of which other sets can be derived from given ones. In particular the axioms of Zermelo [1908] yield such rules; the validity of these still have to be examined.

In order to secure a general principle, from which such rules can be derived, we shall divide the sets and systems of sets into those that are *circle-free* and those that are *circular*. Arbitrarily collected sets can then fail to form a new set only if the definition of the new set contains a non-satisfiable circle: No other obstacle is conceivable. It will be shown that there is always a circle-free set corresponding to each circle-free system. What has to be understood by these concepts must now be investigated more exactly.

For this purpose we exclude as being *circular* all those sets whose transitive closure contains a set that is essential in itself. Certainly all sets which contain themselves, such as the *J*-set and the set of all sets, are thus excluded. The remaining sets form a *transitive system* Σ^*. We restrict the word "set" as it occurs in the following to refer to things of this system. In particular, a set which is essential in another will always be different from it.

It cannot be maintained that all the Σ^* sets are circle-free. In particular a system of sets will also have to be considered as being circular if it does not form a set, as is the case with the system Σ^* itself. Such circular systems can now, as can be shown by examples (cf. concluding remark, §18), also occur as subsystems of elememts of Σ^* sets. These sets, just as much as all those which are essential in them, must then be regarded as circular.

Yet, since one cannot say in general which Σ^* sets can be united into a new set, without first having already grasped the concept "circle-free", and as the "set of all circle-free sets" cannot be represented as a circle-free formation — one comes to recognize that the concept "circle-free" can be obtained in no other way than by one which is circular. One is lead by such deliberations to the following definition which fulfills the requirements:

*A set M of Σ^* is said to be **circle-free** if M together with every set essential in M is independent of the concept "circle-free".*

According to this a Σ^* set is said to be "independent of the concept circle-free" if it can be defined so that the definition always yields the same set, irrespective of which sets are designated as being circle-free.

Every Σ^* set which is not circle-free is said to be *circular*.

§14. Justification of the Definition

With this definition of the concept "circle-free", the question as to which sets are circle-free depends on the manner in which a set can be defined.[6] It must be observed that no requirement is made that the definition of a set must be representable by finitely many words. The concept of finite number has not yet been introduced. It is quite possible for a set to be defined by means of some operation which can be applied more than countably infinitely often. Further, the things under consideration are conceptual and not dependent on language or semantics. The definition must, however, be such that, in every single case, it determines the set unambiguously; otherwise the first axiom would not possess a well determined meaning.

---

[6] In connection with this "concept" it is not a matter of something extra-mathematical, but only of a *correlating* of the designations "circle-free" and "circular" respectively, to the various sets; that is, expressed differently, it is a question of a "function" of these sets.

Since the definition of the concept "circle-free" refers to this concept itself, it is natural to consider whether every set really is unambiguously either circle-free or circular. One can see that this really is the case.

First one sees that the empty set, the "set with no elements", is a circle-free set. By comparison, the "set of all circle-free sets", if it exists, is necessarily circular (cf. proposition 11). Thus some sets are circle-free: There may also be circular sets in Σ^*. It remains a possibility that there are Σ^* sets about which no decision can be made.

Now, let M be an arbitrarily given Σ^* set. Two possiblities arise: First, there exists a definition for M, and for every set essential in M, such that it always yields M even if the class of circle-free sets is modified. This modification could be quite arbitrary, simply classifying some sets as circular and the others as circle-free. The second possibility is that for at least one of the sets essential in M (or M itself), there is no such definition.

In the first case the set M will be classified definitively as circle-free in accordance with the definition. In the second case the set must definitively be classified as circular. There is no ambiguity in the definition here either. In examining whether a set is to be classified as circle-free or not, the classification of circle-free sets should be allowed free variation without regard for the final result. The result is therefore also unambiguous, irrespective of whether the decision turns out one way or the other. From this follows:

Proposition 8. *Every Σ^* set is either circle-free or circular, but not both.*

With this the definition of the concept "circle-free" is logically justified. One can, moreover, also convince oneself that the expression "circle-free" really is valid, i. e., no real circle can arise in a set which is circle-free according to the definition. Using proposition 3 the following holds for Σ^* sets:

Proposition 9. *If M is a circle-free set of Σ^* then every set essential in M (so in particular every element of M) is circle-free and distinct from M.*

One can therefore think of the elements of a circle-free set, M, as being already formed in a circle-free way, prior to the formation of the set M itself. The circle which is contained in the concept "circle-free" has no influence, since the circle-free sets are by definition independent of this concept.

§15. The Formation of Sets

We might expect that an arbitrary class of circle-free sets can always be united into a set. To be sure such a set need not be circle-free in every instance. The following, however, does hold:

Proposition 10. *Every well-defined class of circle-free sets forms a set. This can be either circle-free or circular, but it is distinct from every set which is essential in it.*

What has to be understood by a "well-defined class" appears to be quite clear; however, it is precisely this point which has created difficulties for Zermelo's axiom of separation (see Zermelo [1908, 263], Fraenkel [1922a, 231f.], [1922b], Skolem [1922, 219]). For example, one may designate the class of all sets which do not contain themselves as well-defined, even though it does not form a set. In contrast, the class of all those elements of the unit set which are identical to the set which contains the elements of this class, is obviously not a well-defined class, even though only the one circle-free element of the unit set appears in the definition. One comes to recognize that the definition must be as follows:

A *well-defined* class is understood to be one which is defined *completely, unambiguously, and without inner contradictions*.

Thus, in place of Zermelo's concept "definite", we shall have to put the concept of *well-defined* or *consistent*. A well-defined class of sets is the same as saying that a class satisfies axiom I (§7). The expression "well-defined" class in proposition 6 also has the same significance. In most cases a class of sets should be understood as meaning a well-defined class.

Returning to the proof of proposition 10, let us consider an arbitrary well-defined class of circle-free sets. It is to be shown that there exists a set M which contains the given sets and only these as its elements.

Let Σ_M be the system of sets which are essential in M. This system cannot contain the required set M since otherwise M would be essential in itself and therefore some element of M would be essential in itself too. One thus takes M as a thing not in Σ_M which possesses the relation β to only the given sets. It has to be shown that the assumption that M is a set of Σ^* is consistent.

To begin, consider the hypothesis that M is circle-free. If this hypothesis is correct, then indeed M is a set of Σ^*. But if this hypothesis produces a contradiction, then the assumption that M is circular is consistent. As an example, consider the system Σ_M

containing all the sets possessing some specific property and require M to possess this property too. In this case M would be essential in itself and therefore not circle-free after all.

Now it is to be shown that the assumption that M is circular is consistent. Even though M is now supposed to be circular it cannot be isomorphic to any set of Σ_M, since it would then be essential in itself. The sets of the system Σ_M are all, by proposition 9, circle-free (one could also require this explicitly in the statement of the present proposition). From this definition it cannot be that M, a circular set, belongs to Σ_M; hence M is a new thing. Moreover, the assumption that M is a circular set is in accord with the concept "circular". For, if the definition of M were always to yield the same set regardless of which sets were taken to be circle-free, then it could not, as we supposed, lead to a contradiction if subsequently M itself were to be classified as circle-free.

Therefore, since either the supposition that M is a circle-free set, or the supposition that M is a circular set stands in accord with all the stipulations made, from this proposition 10 is proved.

Because of proposition 8, the class of all circle-free sets is well-defined and therefore, by proposition 10, it forms a set. This set cannot be circle-free, otherwise it would contain itself, hence:

Proposition 11. *The set of all circle-free sets exists and is circular.*

From this the existence of circular sets in the system Σ^* is assured. It might at first appear to be surprising that there are sets which can be defined by nothing more than reference to the concept circle-free, and which therefore cannot be given independently of it by "specifying" its elements. The class of "all" sets, however, is also not given by merely "specifying" elements without using the concept "all". Otherwise the addition of new sets would result in a contradiction.

A *system* of sets is said to be *circle-free* if it contains only circle-free sets and is itself "independent of the concept circle-free" (cf. proposition 12). The elements of a circle-free set always form a circle-free system and conversely, by proposition 10, the sets of a circle-free system form a set which, along with the sets essential in it, must be independent of the concept circle-free and which must, therefore, actually be circle-free. Consequently, circle-free sets correspond to circle-free systems and only to systems that are circle-free. In particular, the following is true:

Proposition 12. *A well-defined class of circle-free sets forms a circle-free set if and only if it is independent of the concept circle-free,* *i. e., if and only if it can be so defined that the definition always yields the same class regardless of which sets are classified as being circle-free.*

Next, let an arbitrary circle-free system Γ be given. Adjoin to this system all of the sets essential in a set of the system. One then obtains a transitive system Σ_Γ, which is still circle-free.

In order to show that the sub-systems of Σ_Γ are also circle-free, consider only the sets of this system, omitting all other sets. Whenever Σ_Γ is given, all the subsystems of Σ_Γ are determined as well; so each of these subsystems must be definable within Σ_Γ, i. e., without reference to sets outside of Σ_Γ. Using such a definition one avoids reference to the concept "circle-free". After all, the sets of Σ_Γ are all circle-free, and Σ_Γ is itself independent of this concept. The condition "circle-free" is always satisfied here, so it can be omitted from any definition of a subsystem which might contain it. Any condition to the effect that a set must be circular would never be satisfied. In this way the definition of the subsystem is rendered independent of the concept "circle-free" and hence is circle-free. In addition, each subsystem of Γ is a circle-free subsystem of Σ_Γ, hence we have:

Proposition 13. *Every subsystem of a circle-free system is circle-free.*

Using proposition 12 simple examples of circle-free sets can easily be constructed, such as the empty set and the unit set. That there do exist circle-free sets which cannot be formed simply through application of proposition 12 is shown by the following example.

We use the abbreviations: $\{ \} = 0$, $\{0\} = 1$, $\{1\} = 2$, etc., and in general, $\{n\} = n + 1$. In this connection we will take for granted the existence of this infinite sequence of circle-free sets. The actual proof will be reserved for the second part [page 161 of this volume]. Here we shall consider the concepts of finite and countably infinite as having already been established.

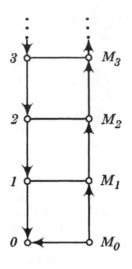

Figure 4

Now add to these sets countably many things: M_0, M_1, M_2, ..., and specify that, for each n, M_n shall possess the relation β to the things n and M_{n+1} so that $M_n = \{n, M_{n+1}\}$ (see Figure 4). The system consisting of the things: 0, 1, 2, ..., M_0, M_1, M_2, ..., then satisfies axioms I and II; it is a complete system and none of its sets are essential in themselves. Further, since these sets are entirely independent of the concept "circle-free" it follows, in particular, that M_0 is a circle-free set.[7]

Circle-free sets in which the empty set is not essential can be formed in a similar way. An example is shown in Figure 5, where each set M_n has n elements.

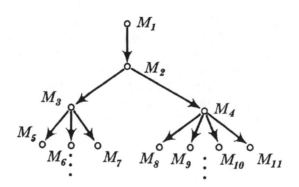

Figure 5

---

[7]   This is an "ensemble extraordinaire" according to Mirimanoff [1917, 42]. — *Editor's note*: Sets having infinitely descending chains under the containment relation were called "ensembles extraordinaires" by Mirimanoff [1917] who first discovered an example of such a set.

§16. *Theorems of Set Existence*

With the help of propositions 12 and 13, we can now establish basic existence theorems which are equivalent to certain of Zermelo's axioms.

As has already been remarked the following holds:

Proposition 14. *There exists a unique set, the "empty set", which possesses no elements. This set is circle-free.*

In addition, one finds that:

Proposition 15. *For any circle-free set A there always exists a circle-free set {A} which contains A as its sole element. For any two circle-free sets A and B there always exists a circle-free set {A, B} which contains the sets A and B, and only these, as its elements.*

Since, by hypothesis, the sets A and B are circle-free, they are also independent of the concept "circle-free". From this it follows directly from proposition 12 that $\{A\}$ and $\{A, B\}$ are circle-free sets.

A set T is, as usual, said to be a *subset* of the set M if every element of T is also an element of M.

If the set M is circle-free then its elements form a circle-free system and by proposition 13 the same is also true of each of these elements. Proposition 12 then yields:

Proposition 16. *Each well-defined class of elements of a circle-free set M forms a circle-free subset of M.*

We consider next the system of all subsets of M. The class of elements of M remains independent of the concept "circle-free". Because a subset T of M can contain only these elements, proposition 16 applies. This shows that T must be a circle-free set and that also, consequently, the fact that T is a subset of M cannot depend upon the concept "circle-free". The class of subsets thus remains independent of this concept and we have:

Proposition 17. *To every circle-free set M there corresponds a circle-free set P(M) (the "power set" of M) which contains all the subsets of M and only these as its elements.*

If M is circle-free then the set $P(M)$ is always different from M; for, M is an element of $P(M)$, but not an element of M. For circular sets, however, the power set can be identical to the original set, as is shown by the example of the set of all sets in the total system Σ.

Those sets which occur as elements of a circle-free set M are essential in M and are therefore circle-free by proposition 9. By means of an argument analoguous to that for the proof of proposition 17 (or alternatively by application of proposition 13 to the circle-free system Σ_M) we find:

Proposition 18. *To every circle-free set M there corresponds a circle-free set $\bigcup M$ (the "union set" of M) which contains all elements of the elements of M and only these.*

§17. The Axiom of Choice

Zermelo's axiom of choice holds for circle-free sets in the following form:

Proposition 19. *Let M be a circle-free set whose elements are pairwise disjoint and different from the empty set. Then, there exists at least one set which has exactly one element in common with each element of M and contains no other elements. Every such set is circle-free and is a subset of $\bigcup M$.*

Since, by hypothesis, every element of M contains at least one element, it can also be assumed without contradiction that in each element A of M some fixed element a of A will be distinguished and correlated with the element A.[8] That the element a is distinguishable from the other elements of A follows directly from axiom II; otherwise they would be isomorphic. The uniqueness of each set gives any choice set a distinct character. That much is true whether or not the sets of M are mutually disjoint. By "choosing" elements independently of each other we are certain that no unsatisfiable circle can arise.

The choice class consists of elements of $\bigcup M$. By proposition 18, $\bigcup M$ is a circle-free set, so by proposition 16 the choice class is a circle-free set too. A set found as above will possess the required properties. Any other set which has exactly one element in common with each element of M and no elements other than these is a subset of $\bigcup M$ and so, by proposition 16, is circle-free as well.

Using proposition 19, the axiom of choice is not an independent axiom nor an arbitrary stipulation but a provable proposition, i. e., it

---

8   The correspondence is necessary only for proposition 20.

is logically deducible from the axioms. In fact, the negation of this statement would contradict axiom III. Since the system Σ^* and, by proposition 8, the system of all circle-free sets as well, are both unambiguously determined, one can see immediately that the proposition can only be either true or false.

The proof given for proposition 19 is independent of the distinction between finite and infinite sets, which has not been given here. There does, however, still remain the question as to whether the proposition can be proved in an essentially different way, or whether one might show that this is not possible.

The axiom of choice has assumed a position of special interest so it may well be appropriate to add a few remarks concerning its place within the axiom system of Zermelo.

If one assigns a restricted meaning to the concept "definite" in Zermelo's axiom of separation then, as Fraenkel [1922b] has shown, the axiom of choice is independent of the remaining axioms, because there exist sets whose existence depends upon the axiom of choice.

Attention must be paid to the fact that the sets obtained using the axiom of choice must be subsets of the set $\bigcup M$. They are not singled out by a definite property however. It follows that the concept of "all" subsets of a set is *not uniquely determined* by the axiom of separation. An "axiom of restriction" (Fraenkel [1922a]) does not help either bring about a unique determination, since, under circumstances in which several possible sets are given by the axiom of choice, an axiom of restriction would admit only a single, arbitrarily selected one of them.[9]

If the concept "definite" in the axiom of separation is understood so broadly that it yields *all* consistent subsets of a set, then the axiom of choice would not really serve the purpose of determining new sets but would at most tend to exclude certain collections from being sets.

Thus, in the case that there is an "axiom of restriction" that admits only those things allowed by the axiom, then the axiom of choice is either superfluous or untrue, i. e., it is either a consequence of the other axioms or it contradicts them.

In a certain sense these remarks furnish an explanation for the distrust which has arisen around the *axiom* of choice (cf. also §18).

That it is necessary for the elements of M to be disjoint in proposition 19 is shown by the example of the set: $\{\{a\}, \{b\}, \{a, b\}\}$ in which a and b denote the empty set and unit set respectively. As Zermelo [1908, 274] has shown a general principle of choice that is

---

[9] *Editor's Note*: Fraenkel's "axiom of restriction" limits the universe of sets to those explicitly guaranteed by the other axioms.

still valid, even when this condition is not satisfied, can be found. It is:

Proposition 20. *Let M be a given circle-free set which does not contain the empty set as an element. Then there exists at least one circle-free set P which associates with each element A of M an element a of A.*

As in the proof of proposition 19 one can assume without contradiction that to each element A of M there corresponds a definite element a of A. Now, using proposition 15, combine each element A of M with the corresponding set a to form a set $\{A, a\}$. The sets which arise in this way are all subsets of the set $\cup\{M, \cup M\}$ and are consequently elements of the circle-free set $P(\cup\{M, \cup M\})$. So, by proposition 16, they form a circle-free subset of this latter set.

The set $R = \{ ..., \{A, a\}, ...\}$ obtained in this way unambiguously associates an element a of A with each subset A of M as required; for in the system Σ^*, $A \ \beta \ a$ is never satisfied at the same time as $a \ \beta \ A$, therefore no ambiguity can arise.

§18. Further Remarks

Propositions 5 and 14 to 19 correspond to Zermelo's [1908] axioms I to VI and are valid for arbitrary circle-free sets.

Proposition 5, which corresponds to the axiom of extensionality, is valid also for circular sets, but it is not always sufficient for securing the identity of sets in general, as has already been said in §8. That the axiom of extensionality is inadequate even for circle-free sets can be shown using the example at the end of §15. If in this example the sets $M_0, M_1, M_2, ...$ are replaced by $M_0', M_1', M_2', ...,$ which satisfy the equations $M_n' = \{n, M'_{n+1}\}$ then it cannot be proved that $M_0 = M_0'$ by using proposition 5 alone in place of axiom II (§5).

The consistent existence of the sequence of natural numbers can be derived without essential difficulty from what has already been established (cf. Dedekind [1918]), but it is not self-evident, since the usual definition of the natural numbers, and that of the transfinite ordinal numbers, is of a circular nature. Both will therefore be treated in the second part of this investigation.[2]

With the existence of a circle-free set which corresponds to the natural numbers, the seventh axiom of Zermelo (axiom of infinity) will then have been proved also. Thus, since all the axioms of Zermelo are satisfied in the domain of circle-free sets, the consistency of this axiom system follows immediately. Therefore, all results provable from these axioms, hold for circle-free sets. By using proposition 12, however, it is still possible to form, in a simple way, sets which cannot be obtained from Zermelo's axioms alone (cf. Fraenkel [1922a, 230ff.]; Skolem [1922, 225]).

In particular, because of proposition 16, the theorem of Cantor stating that the set of all subsets of a set has a greater cardinal number than the set itself, holds for circle-free sets. From this it follows also that the set of all circle-free sets has a greater cardinal number than every other circle-free set. By proposition 16, this set must contain the subsets of its elements. Whether, in general, each circular set whose elements are circle-free possesses a greater cardinal number than each circle-free set is a matter that still requires investigation.

The set of all circle-free sets is an *infinite* set in the sense of Dedekind [1918], since it contains the empty set and, by proposition 15, for every set A it must also contain the set $\{A\}$; it is thus equivalent to a proper subset of itself.

Since by proposition 10, each class of circle-free sets forms a set, the set of all sets in the total system Σ, which indeed contains all these sets, must have a greater cardinal number than the set of all circle-free sets. Thus the existence of large cardinal numbers is demonstrated in an absolute sense.

For circular sets, in particular for the set of all sets in the total system Σ, the axioms of Zermelo are no longer all satisfied; for example, the sets which do not contain themselves are all elements of the set of all sets, but do not form a subset of it.

It can be shown that the union set axiom and the axiom of choice also are not satisfied in the system Σ; but as complete proofs using circular sets are difficult the following will show that this contention is at least plausible.

Let N be a set that contains all sets that do not contain themselves, but excluding N itself and nothing more. In the formation of this set there is no obvious contradiction and it is therefore plausible to suppose that it does exist. Now form the set $\{N, \{N\}\}$ in which again no contradiction is recognizable. One cannot, however, form the union set for these two sets; for, it would have to contain all sets which do not contain themselves and these alone.

Consider further the set which contains all and only those sets that contain as their sole element a single set which does not contain itself. The axiom of choice is not satisfied for this apparently consistent set. The choice itself is possible in an unambiguous way, since each set from which the choice is to be made contains exactly one element; but the chosen elements cannot be united into a set.

If one defines the product of a set as the set of all sets determined by the axiom of choice, then, in this case, the product is empty without one of the factors being empty.

The same example shows moreover that one need not always obtain a set by replacing elements of a set with other elements. Namely, if in the set just considered, each element is replaced by the one set which is contained in it then one gets, once again, a class of sets which does not form a set.

In the second part [page 161 of this volume] similar examples will be constructed for the system Σ^*.

Additional Remark

For the avoidance of apparent objections it should still be remarked that: If, in some system of things, a relation is defined in terms of *another* relation which is assumed to be already given in the same system, then the former is said to be a *derived* relation, in contrast to the latter which is the *original*, or *primitive* relation and which is given first. The relation β is always to be taken to be a *primitive* relation.

First published as: "Die Existenz der Zahlenreihe und des Kontinuum", *Commentarii Mathematici Helvetici* 5 (1933), 88–94.

The Existence of the Number Numbers and the Continuum

§1. Preliminary Remarks

The absolute consistency of the sequence of natural numbers and of the continuum will be proven here.

The proof is based on earlier work (Finsler [1926b]) in which a consistent axiom system for set theory was established and from which some propositions were derived. This work was dismissed by some (Baer [1928a]). But valid objections concerning essential points are unknown to me; in fact I claim them to be impossible. The following short remarks may suffice to make my standpoint clear.

1. Someone who cannot resolve the paradoxes and antinomies of set theory and logic (or who holds them to be insoluble) is not in a position to criticize such a theory. An unresolved antinomy could be used to prove or refute anything.

2. Someone who can resolve the antinomies in the right way knows that pure logic represents a sure ground upon which to build. To consider a formal system as being "more exact" than pure logic is an error: Formal statements alone do not suffice for removing the antinomies (cf. Frege [1893], [1903, Nachwort]). This can only be done by thoughts which are based on pure logic standing above formal representations.

3. To accept finite but unlimited induction as being given would be a petitio principii: But to admit only the finite would unduly limit mathematics. Mathematics is something more than a handicraft or a game like chess. The possibility of transfinite contradictions must be excluded too.

§2. A Proof is Necessary

The natural numbers are readily thought of as something given directly. If one does this only provisionally, then it is certainly justified. If, however, one proceeds with the utmost care, then one will not admit anything that is unproven. It may be said that small numbers are given to us directly, when we can survey them in complete detail. But for large numbers it becomes more doubtful. The natural numbers as a whole certainly are not given to us directly.

Evidence for this is to be found in the fact that there were, and perhaps still are, many mathematicians who completely reject the existence of the natural numbers as an entire system with infinitely many elements. The reasons for this rejection have to be investigated; one would not want to simply reject something without proof. Furthermore, before attempting to prove the consistency of the natural numbers, one would need to be clear about what is to be proven and where the difficulties lie.

The expressions "natural number" or "sequence of numbers" mean nothing by themselves; one has to say what is meant by them. An exact definition is required before one can prove something. The properties that we derive must follow from the definition. In the case of the natural numbers there are two definitions to be considered: the generative definition and the axiomatic one.

(a) The *generative definition* starts from a given, initial number 1 (whether one begins with 0 or 1 is irrelevant). Next, introduce an operation +1 which yields, for each number already found, a new number different from all the others previously obtained. In this way one obtains 1+1, 1+1+1, 1+1+1+1, etc. which can then be called 2, 3, 4, etc.

Is it now possible to speak of the totality of all numbers? If we accept that there exists a given, well-defined system of things from which the individual members of the sequence are taken, then we could define the natural numbers as the totality of things in the system obtained by proceeding from the number (the thing) 1 with arbitrarily many applications of the operation +1.

If, however, one cannot assume that such a system of things is given, then this conclusion is not permissible. The sequence of natural numbers would only be in the process of "becoming". Each number would then be newly created only on the basis of the earlier ones. We would not see, to begin with, whether or not this process would finally come to an end or not.

One might think: Of course this process does not end. This thought overlooks an important point: There has been nothing to justify the assumption that a successor can really be found for every natural number.

The existence of an immediate successor to every number appears to be required in the definition. But a definition alone cannot secure the existence of something. Is one to be allowed to postulate the existence of something which could even carry a contradiction hidden within it and therefore not exist at all? That is exactly how the antinomies arise. We cannot create something logically inconsistent, and therefore non-existent, by an arbitrary act.

Why do we believe that every number has a successor? Is it because for every known number, there is a larger one? The numbers that can actually be given to us form only an infinitesimal portion of the natural numbers; so this fact cannot serve for a proof.

One might say that there are no reasons to be advanced against the existence of an immediate successor to each natural number, therefore there could be no contradiction from this assumption.

But there are such reasons, one can reply. The definition that generates the natural numbers is circular; and circular definitions are not always satisfiable.

The circle becomes apparent when the definition is given a more exact form. The generative definition is essentially a rule for construction which can be given in the following form.

Construction A: Begin with the number 1 and place a new number after each number which results from construction A.

In the first attempt to define the separate natural numbers one had to put a special symbol, +1, after each number obtained by this very rule.

This rule, or construction, therefore explicitly refers to itself. Hence it is circular. The hazards of this circle can be recognized in a different but analogous rule of construction.

Construction B: Begin with the number 1 and place a new number after each *sequence of numbers* which results from construction *B*.

The construction *B* is certainly not satisfiable in every instance.[1] An attempt to do so would reproduce the antinomy of Burali-Forti. Therefore, it is not self-evident that construction A is satisfiable either. *The statement that every number has a successor requires proof.*

(b) In the *axiomatic definition* the natural numbers form a system of things which satisfy certain axioms. These axioms can be given the following form, due to Peano.

---

[1]   Thereby it is of course assumed that one can also conceive of each sequence of numbers (e. g. the sequence of all natural numbers) as being a whole. This will be shown for the natural numbers in paragraph 3 and is in general possible on the basis of the system of all sets, which is defined in Finsler [1926b].

(I)   1 is a number.

(II)   If n is a number then so is $n + 1$.

(III)   If m and n are numbers such that $m + 1 = n + 1$,
        then $m = n$.

(IV)   For every number n, $n + 1 \neq n$.

(V)   A proposition that holds for the number 1 and that holds
        for $n + 1$ whenever it holds for n, holds for every number.

An axiom system of this kind has the advantage that it specifies, separately and completely, each of the requirements which are placed on the concept of the natural numbers and which are necessary in the definition. This provides a foundation upon which arithmetic can be built. The question of the security of this foundation remains: Does the axiom system have contradictions or not?

To settle the matter one could attempt to show that it is impossible to arrive at a contradiction "in finitely many steps". Such a proof would be important, but it would not be decisive.

That is to say, if one assumes that there is a natural number having no successor, then there would also be a limit to the length of a proof. An inconsistency might not be revealed in those proofs that lie within the permitted limits. Once again, there would be no guarantee that every natural number has a successor, even though the negation of this statement stands in overt contradiction to the axiom system. *This* inconsistency would not be formalizable and hence not susceptible to analysis by formal methods.

However, if one can find a system of things whose existence is consistent and which satisfies the axioms, then the axiom system is shown to be consistent in an absolute sense. We have already seen that the generative definition does not provide such a system. But a system of this kind can be obtained from set theory.

Just as the consistency of geometry can be proven from arithmetic, the consistency of arithmetic can be proved from that of set theory. For set theory one must use a different method; one has to base the argument on purely logical grounds as was shown in Finsler [1926b].

The consistent existence of the continuum will follow from that of the natural numbers, provided that the operations of set theory allow one to form arbitrary subsets of the natural numbers.

§3. The Proof

By "sets" we mean ideal things which are related to one another by means of a fixed relation: "containment". Collections of sets are called "classes". There does not need to be a set corresponding to each class. The class of all sets determined by the three axioms is denoted by Σ. For some of the propositions as well as for the concept "circle-free" one must refer to Finsler [1926b].

Consider those classes Γ which contain the empty set, and contain $\{M\}$ whenever they contain the set M, for each M such that $\{M\}$ is also a set. There is such a class, namely Σ itself. Form the intersection of these classes; it will contain those sets that are in every such Γ. Let Δ be this intersection. The empty set is in Δ, and for each M of Δ, $\{M\}$ is a member of Δ whenever $\{M\}$ exists. Moreover, the *induction principle* holds for this class, Δ, as shall be shown.

Let ϕ be a statement which holds for the empty set, and holds for the set $\{M\}$ whenever it holds for an M such that $\{M\}$ exists. It is to be shown that ϕ holds for every set of Δ.

For the proof, consider the class of all sets for which the the statement holds. This class contains the empty set and contains $\{M\}$ for each of its members M, provided that $\{M\}$ is indeed a set. Thus it is one of the classes Γ, and all the members of Δ must belong to it. Thus the statement ϕ holds for each member of Δ. This provides an induction principle for Δ.

In Finsler [1926b] it was shown that the empty set is circle-free and that, in proposition 15 of that paper, the set $\{M\}$ is circle free whenever M is. Thus, using the principle of induction established above, we have that Δ really does contain $\{M\}$ whenever it contains M.

If the set $\{M\}$ is identical with $\{N\}$ then M is identical to N by the Axiom of Identity. None of these sets can be equal to the empty set, since all the sets of the form $\{M\}$ contain an element.

This shows that Peano's axioms are all satisfied with the empty set in place of 1 and $\{M\}$ in place of $n + 1$, whenever n is a name for the set M.

In the definition of the class Δ there was no reference to the concept "circle-free" so Δ is independent of this concept. Consequently there is a circle-free set T which contains exactly the members of the class Δ as its elements. This follows directly from proposition 12 of Finsler [1926b]. This shows the axiom of infinity holds in the circle-free sets, a result which was omitted from the paper just mentioned. Accordingly, all of Zermelo's axioms hold in the circle-free sets. It follows that Zermelo's axiom system is consistent.

Using proposition 17 of the same paper there exists a circle-free set which contains exactly the subsets of T as its elements. This is well known to be equivalent to the existence of the continuum.

§4. Concluding Remarks

One could ask whether the existence of the natural numbers and of the continuum could be proved without use of the concept "circle-free".

This might be possible for the natural numbers. One can show, although not in a simple way, that for a set M which is not circle-free, the set $\{M\}$ exists; circular definitions might arise however.

For a proof that the class of all sets contains a set equivalent to the continuum, the concept "circle-free" or something equivalent to it, probably cannot be avoided in any case.

If the natural numbers are given one could take the collection of all sub*classes* of the natural numbers as the continuum. This would constitute a collection of a higher order. In forming functions there would appear collections of even higher order. This would not only be inconvenient but could lead to difficulties. In any case, one could not obtain a general set theory in this way.

The advantage of the method employed here is that the first order objects, the sets, already emcompass the whole of set theory. The second order objects, the classes, and the third order, all classes possessing some property, only appear in the foundations of the subject. Once the foundations of analysis are established, there is no further need to concern oneself with objects of higher type.

Lecture at the Mathematical Colloquium, University of Zürich, January 1939. First published as: "A propos de la discussion sur les fondement des mathématiques" in: F. Gonseth (ed.): *Les entretiens de Zurich sur les fondements et la méthode des sciences mathématiques, 6–9 décembre 1938*; Zurich: Leemann 1941, 162–180 (MR 2, 339).

Concerning a Discussion
On the Foundations of Mathematics

An International Congress on the foundations and methods of mathematics was held here in Zürich a month ago, in December. The credit for this belongs to F. Gonseth who inspired and led the conference with the assistance of the Institut International de Cooperation Intellectuelle.

The aim of this Congress was not to bring the discussion concerning the foundations of mathematics to a definite conclusion in just a few days. Rather the intent was to lead this discussion along a useful path. Toward the end of the conference several questions arose, which in my opinion are very serious and which could not be resolved. It seems reasonable to me to take up these questions again and pursue the discussion further. There is also this additional reason: If one knows that a train is on the wrong track, one has a duty to stop the train, if it is possible, in order to prevent disaster. I judge that it is the same in mathematics; and this is why I am speaking on the topic.

I shall consider a certain problem in set theory: the *nature of the axiom of choice*.

Only those things that arose during the Congress which are necessary for the understanding of what follows will be repeated here. I claimed there that the axiom of choice is not truely an axiom at all within a well-defined and complete system of sets, but a proposition which could be true or false like any other. I would like to make this claim more precise and at the same time limit it somewhat.

To begin, here is the content of the axiom in question. Let M be a collection of non-empty pairwise disjoint sets. The members A, B, C, ... of M can have any power whatever. There is at least one element a in A, B has at least one element b, and so on. Because the sets have no elements in common, their elements must be different from one another. Zermelo's axiom of choice asserts that for each such set M there is a choice set N having exactly one element in common with each set in M. This means that N has exactly one element a from A, a unique element b from B, and so on.

The choice set N is a subset of the set $\bigcup M$ which contains all the elements of A, B, C, ... Were A, B, C, ... to contain only one element each, a, b, c, ... respectively, then one would have that N is idential to $\bigcup M$. In this case it is not necessary to use the axiom of choice to construct the choice set N. In certain other cases one can use the axiom of separation to define the *choice set* as a subset of $\bigcup M$. The question now arises: Under what circumstances is this impossible? That is to say, when is the axiom of choice essential? At the same time, we must ask whether the axiom of choice is really satisfied.

If one proposes some system of axioms for set theory, one can envisage three possible cases for the axiom of choice. The first is that there is a set M, as above, for which there is no choice set: The axiom of choice would then be *false* or the axiom system would be contradictory. In the second case, one could always find a choice set N for a collection M using the other axioms: The axiom of choice would then be *superfluous*. The third case still remains: the case in which the axiom of choice is genuinely necessary to determine some choice set.

This last case only arises when the axiom of separation fails to provide all subsets of some set: In that case it would obtain all the choice sets too, if there are any. It is possible however that the axiom of separation is conceived in so narrow a way that it only provides those subsets that are constructed in this or that manner. The remaining subsets, for which one requires the axiom of choice, cannot be constructed in the same sense; there is no rule for construction included in the axiom of choice. If one admits into mathematics only those things that are explicitly constructible, one would reject the axiom of choice on principle, *a priori*.

Suppose, on the other hand, that we do not regard sets as being given only by specific constructions. In this case the axiom of choice is generally necessary unless there is a unique choice set. A choice set that is unique could be obtained easily. If there were no choice set then the axiom would be false. So the axiom would give at least one of the available choice sets. Thus it is possible, or at least compatible with the present axiom systems, to take only one of the choice sets. So far there is no indication, however, of which choice set is preferred. The sets provided by the axiom of choice are unspecified. The axiom is *vague* because it does not precisely specify which sets it actually provides. In a categorical system this could not happen.

This does not mean, however, that there is no categorical axiom system, one whose universe of sets could not be affected by dropping the axiom of choice (such modifications seem to be possible only in artificial systems).

The axiom of choice taken as a proposition can only be true or false, as with any proposition, in a categorical system not including this axiom. In particular this is the case for an absolute set theory that admits all sets except those that would produce a contradiction.

One might mention another interpretation of the axiom of choice in which choice sets are not absolutely determined. It asserts that among the choice sets that are assumed to exist, one of them (determined or freely chosen) is provided. In this case, not only is the name "axiom" a bad choice but the very formulation itself would require modification. In any case, this could not have the same status as the other axioms of set theory.

*   *   *

Now the question arises as to whether the axiom of choice is true or false in absolute set theory. A communication from K. Gödel was read at the Congress saying that Mr. Gödel succeeded in demonstrating that the axiom of choice is *free of contradiction*. I would like to offer a contrasting communication according to which the axiom of choice is *false*. This is the case not merely for specially constructed systems: But it is false in a very important and general case.

These two results, that of Gödel and the one I announce, are not really contradictory. It is quite possible to prove consistency (that is, the absence of formal contradiction) for propositions that are evidently false. I cannot accept that it is permissible to make an off hand classification of a false proposition as being free of contradiction. Nor do I see why years of labor should be devoted to proving the consistency of propositions and theories even if, having achieved success, it can still turn out that they are evidently false.

This *evidence* is not to be denied: It is too simple. I will give it by an example which I have long ago introduced (Finsler [1926a]); it is still of value today.

Suppose we are given a formal system, anything which is neither inconsistent nor too restricted will do. Certain statements are provable in this system. In any comprehensible formal system there are only two alternatives for any given formula: It has a proof in the system or it does not.

Now consider an arbitrary infinite sequence formed, for instance, using the digits 0 and 1; the sequence $001100110011\cdots$ can serve as an example. One could ask whether or not the digit 1 appears infinitely many times. For some sequences one can formally prove which of these possibilities holds: The sequences $1111\cdots$ and $101101110\cdots$ are of this kind. Each of these proofs contains only finitely many symbols; hence they can be enumerated. To each proof there corresponds a unique binary sequence, the one treated in the proof. The binary sequences themselves can be enumerated by the enumeration of proofs, though a sequence could perhaps appear more than once.

Now we construct a new sequence, a, using Cantor's diagonal method in which the nth digit differs from the nth digit of the nth sequence in our enumeration. That is, one replaces 0 in that position with 1, and vice versa. Finally, consider the proposition: *In a the number 0 does not appear an infinite number of times.*

If the formal system does not permit a proof contradicting this proposition, then the proposition is formally consistent. So a would be one of the enumerated sequences: But this is impossible. The proposition a is, however, clearly false. To see this consider the sequence $1111\cdots$, for which we can find more and more complicated, though still finite, proofs showing that this sequence does not contain 0 infinitely many times. For example, 0 does not appear after the first place, nor after the second place, nor after the third, nor after the rth place. For each such proof the sequence a will contain a 0. Hence 0 occurs infinitely often in a.

This argument shows that the proposition given above is formally consistent yet evidently false. Such a flagrant contradiction cannot be simply branded "non-existent".

Further, it follows that a consistency proof for some theory does not guarantee that the theory is free of contradiction. I believe that this is an important objection to formal proof theory. It was only when this objection was made arithmetical and hence more limited in scope (known as Gödel's theorem) that proof theory took it seriously and was transformed significantly. There is a question of principle here; does one objectively consider criticisms directly where they arise or does one act when forced by outside events? The essential content of my objection has not yet been taken seriously into account.

Without knowing the extent to which Gödel's proof has been realized, I cannot actually claim that something false has been shown consistent.[1] In analysis the axiom of choice is true, not false at all. This result is not new. Several years ago I showed the validity of the axiom of choice in analysis (Finsler [1926b], [1933]). Formal consistency of a proposition follows from its general validity. It also follows that the continuum can be well-ordered, but this does not provide an effective well-ordering. There always exists a difference between those things provable without the axiom of choice and those things which require it.

---

[1] I have recently seen Gödel's note [1938]. Gödel only establishes the *conditional* proposition: if von Neumann set theory is consistent without the axiom of choice, it is consistent also with this axiom. My objection is one of principle and is not altered by this.

On the other hand, Gödel's proof concerns set theory; it is just in this domain, however, where the axiom of choice as originally formulated by Zermelo is false. I shall give an example soon. First, it is necessary to discuss another question that arose during the Congress, that of the *status of arguments that are not formalized*.

Think of the research that has been undertaken to show that analysis is consistent. I asked myself what had been accomplished by these years of extremely complicated and laborious research. P. Bernays himself acknowledges that there has been little progress towards a real understanding of the actual consistency of analysis; the formal research does not guarantee that actual contradictions are not present. Yet one can obtain the absolute consistency of analysis by a few lines of argument that are plainly correct and not very difficult. I ask again: Which is better?

On the other hand, absolute consistency insures formal consistency (and not conversely). As a consequence, all the long and painful research undertaken to show formal consistency is superfluous in the end: The result has long ago been obtained. This last objection still holds. The result was obtained many years ago but nobody paid any attention to it.

It is the same in set theory. The axiom of choice is true in certain partial domains of set theory and hence is consistent. But it is false in the entire universe of that theory. To be fully secure, it is surely necessary first of all to settle on the nature and objects of set theory itself. I would like to put forth how this can be done without unduly restricting the concepts of set and cardinality.

Some limitations on sets are necessary; if a set is to be well-defined, then it cannot be left unstated what its elements are to be. Objects from outside of mathematics are not permitted as elements. Even mathematical objects are not to be taken as properly defined unless they have a basis in set theory. One leaves aside all but the pure sets themselves; sets only have other sets as members.

In order to define the sets exactly, it is best to fix them axiomatically. Sets are just "things" between which there is a relation satisfying the axioms. One could take for this relation the membership relation $\in$, where $a \in M$ when a is contained in M as an element. For a reason which will soon be clear, I use the converse relation "containment": $M \beta a$, that is, M contains a as an element.

The totality of sets is now given by the following *three axioms* (Finsler [1926b]).

The *first* asserts that it is always uniquely determined for any two sets M and N whether or not $M \beta N$. In other words, for any set M, it is always determined which sets hold the relation β to it. This means that the elements of a set are determined. The elements of a set are always determined; but this does not mean that simply determining elements insures that there is a set that contains them. This is precisely the reason for choosing the relation β in preference to ∈.

The *second* axiom says that isomorphic sets are identical. Sets which cannot be distinguished from other sets will not be admitted.

The *third* axiom is that of completeness. One requires that all sets are to be admitted that satisfy the other axioms.

An objection has been raised against this axiom of completeness in Baer [1928a] that it is unsatisfiable, and does not prevent the whole system of sets satisfying the first axioms from being expanded.

This objection is mistaken: It does show however that one must be clear about the antinomies to be able to judge these things. With an unresolved antinomy one could, of course, refute or prove anything, but this is not the way of scientific progress. Resolving the antinomies is therefore absolutely necessary (Finsler [1925], [1927b]).

Even today one hears the opinion that the antinomies arise from considering infinite totalities as complete wholes and that this should not be permitted, or that the formation of uncountable cardinalities should not be permitted. But why should this be forbidden? We are not given an objective reason; and besides, the antinomies arise from a different direction. It is perfectly permissible that one should operate with any actual infinite number and construct powers however large. Mathematics permits everything that is free of contradiction. This is the only mistake that one can commit; if one avoids it the antinomies disappear. Set theory remains intact with its powers, higher and still higher.

That these powers exist is a fact, but a fact, it is true, which is not to be admitted without some consideration. One first must make sure that no contradiction lies hidden within their definitions, and this is not easy. But this difficulty does not start with the uncountable; it is already present with the sequence of natural numbers. For the sequence of natural numbers to be extended arbitrarily it is necessary to suppose the existence of infinitely many things and thus to pass beyond direct experience. The usual definition of the natural numbers contains a vicious circle and only a profound investigation shows that there is no contradiction, that is, that the sequence of natural numbers exists (Finsler [1926b] and [1933]).

What about the system of all sets just defined? Can one enlarge it? The system containing the natural numbers, the continuum, and the subsets of the continuum can always be enlarged. But if we have all sets, it is different. If I say that I have taken all, and then I take one more, I have contradicted myself. That is not permitted. If I say only that I am taking all sets, there is no contradiction. I must not then say that I am going to take still more. The system of all sets cannot be enlarged.

It is necessary to take up another objection to this theory which is also based on a misunderstanding.

I have spoken of "systems" of sets, the system of sets, certain systems of subsets, and so on. What is a system? A system is a collection of sets. The sets are "things" satisfying the axioms. One can collect things into a system.

Now, in the course of the discussion during the Congress, someone remarked that one ought not use a vague, intuitive, naive, notion of a set along with an exact axiomatic conception. The idea of a system is an intuitive conception of a set to be sure, but it is free of vagueness and imprecision. On the contrary, if sets are well-defined, so are systems of sets. On the basis of concepts introduced using the axiomatic method, one can follow with other precise concepts. For example, if the natural numbers are given by the axiomatic method, one can, on that basis, introduce the rationals which are not less precisely defined though they are not given axiomatically themselves. It is the same with these systems of sets.

One ought, it is true, assure oneself that there is no new clash with the paradoxes of set theory. But that is not the case. These difficulties only arise when one considers sets of any sets whatever, not with sets of other objects such as point sets. Sets of objects can be formed straight away. Difficulties could arise if one were to form systems of systems in an unrestricted way. But that is not the case here because one only considers systems of sets; no reflexive reasoning appears. It is these reflexive arguments which most easily allow contradictions to slip in; so one must be especially attentive with them. But if the sets have already been defined, then reflexivity does not enter into the formation of systems of sets. There is for example no *set* within the system of all sets containing those sets that are not elements of themselves. But these sets do form a well-defined *system*. Hence there are systems that do not correspond to sets. Instead of antinomies we have theorems, precise and reliable.

It is necessary to add a remark concerning a matter of principle: I do not hold that it is the object of the foundations of mathematics, as it is sometimes presented, to reduce arguments that are more or less doubtful to ones that are merely less doubtful. Or, as one also hears, to replace arguments which are more or less sure by ones that are more sure. It is really not a question of arguments being "more sure" or "less sure", but whether they are true or false. It does not help the situation to mix good arguments together with doubtful ones and consider wrong arguments, even actual errors, as being beyond doubt - as has actually happened.

I shall now construct an *example for which the axiom of choice is false*.

Consider the sequence of finite and transfinite ordinal numbers. One can define them, following Zermelo, as sets. The first is the empty set and the others are identical with the set of their predecessors. That is,

$$0 = \{\,\}, \ 1 = \{0\}, \ 2 = \{0, 1\}, \ 3 = \{0, 1, 2\}, \ ..., \ \omega = \{0, 1, 2, 3, ...\}, \ ... \ .$$

One needs to require that no ordinal contains itself but only contains its predecessors, those strictly smaller.

The set of all these ordinal numbers does not exist. Otherwise the Burali-Forti antinomy would follow. Now consider the sets that contain a single ordinal number, δ. That is:

$$\{0\}\,, \{1\}\,, \{2\}\,, \ ..., \ \{\omega\}, \ ..., \ ... \ .$$

Except for the first one, these sets M_δ are not ordinal numbers because they contain only one element, whereas ordinals contain all their predecessors. Let M be the collection of all these sets M_δ.

The definition of this collection, M, does not involve any contradiction. To show that M is a set it suffices to show that the relation β is uniquely determined; and this is the case. The class M does not contain itself. Nor is it contained in any one of its elements, since it is not an ordinal number; ordinal numbers contain only ordinals as elements while M contains other things too. Therefore by a general proposition (Finsler [1926b], Theorem 6) M must exist.

The axiom of choice fails for this set. A choice set would have to contain exactly the ordinal numbers; and that is impossible. The actual choice is possible and even unique; but the axiom requiring that the selection forms a set cannot hold. This example does not stand against the axiom of choice as such, but only against the axiom of choice in so far as it requires the existence of a set.

It would be interesting to know how far these considerations could be applied to other axiom systems for set theory. The research of Gödel does not apply to the set theory we have just considered, but rather to more restricted systems which have been formalized in one way or another. The same example could be introduced into a more restricted theory provided one were allowed to form the sets of M and all the ordinal numbers.

The result is the following: If one claims that the axiom of choice is compatible with certain other axioms of set theory, in the sense that one can find systems in which all the other axioms are satisfied, the proposition is true but it is not new. If one claims, on the other hand, that in every consistent set theory one can employ the axiom of choice to form new sets, then this claim is false, even if one could prove it to be free of formal contradiction (cf. footnote 1).

Additional Notes

I shall now respond briefly to several objections presented during the same mathematical colloquium, mostly by Paul Bernays.

"The theory rests on a certain philosophical position."

In effect, this position consists simply of the following: that one knows in principle the difference between true and false. And that in the general theory of sets, as in arithmetic, every unambiguous question has a unique answer. We could earn a living without such a philosophical position but the practice of science would be impossible without it.

"The argument leading to the example of a formally consistent proposition which is false must be rejected because it is associated with Richard's paradox."

Is one to be convicted on account of his associations when he is otherwise innocent? The accused must be tried on evidence of a mistake that he has himself made. Where is the mistake in the preceeding argument? For my claim, it is of little significance that the proposition in question can be expressed in this or that formalism: We can easily imagine a suitable formalism. We could even use natural language. We could employ the same method to treat Richard's paradox (Finsler [1926a], [1927b], [1927c]). In any case one has to make some effort to think about these things and then the result becomes evident.

"What does evident mean?"

A thing is evident when, after sufficient reflection, one sees that it holds, and cannot fail. It is possible for a thing to be evident to one person and not to another when he has not attended to it sufficiently. It is impossible for two contradictory facts to be evident. A real contradiction, not merely an error, would make all science impossible. One must decide case by case whether an argument is evident or not, without giving an explicit general rule.

"Is it sufficient to fix the concept of a set by these three axioms?"

The term "axiom" used to designate the conditions which establish a theory is perhaps misleading sometimes. The word "postulate" might be adequate, but it is little used now. Apart from this nuance, it is necessary to fix the notion of a set if one wants to treat set theory as an exact science. This cannot be done by explicit definitions because there are no things already present upon which sets can be based or to which they can be reduced. Hence the definition must be implicit. This means that certain notions are given by their characteristic properties and others are deduced from those.

"Objections against the axiom of completeness have already been raised in various ways."

All these objections are erroneous, just as all the attempts to square the circle must run aground. Certainly it is possible to construct systems larger than the system of all sets. For example, one could adjoin a new *object* for each class that does not form a set. But the new objects do not have the original relation β to the other sets but another relation $\mu(\beta)$ depending on β. Hence they do not form sets because in the axioms there is only one fundamental relation; it is unique and hence does not depend on any other fundamental concept. If one wants, in opposition to this, to designate the new things as sets in the sense of our axioms, in other words to regard $\mu(\beta)$ as a primary relation, then a contradiction results. So one has made a mistake which is not permissible.

"If one thinks of sets as given one by one, why is it impossible to adjoin yet another one?"

This could be done as long as one did not employ the concept "all". One cannot speak of all sets and then adjoin another without contradiction.

"Why not take a universal totality which contains everything imaginable including the new things in question?"

This is possible but unnecessary. Of course, this cannot be one of the sets given by the axioms. The latter system is so encompassing that it contains all of analysis and all formal set theories known until now; they represent only an extremely small part of this system. Besides it is given simply and uniquely. Surely this is large enough.

"You did not mention the division of sets into circle free and circular."

This distinction (Finsler [1926b], Burckhardt [1939]) is of the greatest importance for the further development of set theory and, above all, for the foundations of arithmetic and analysis. For other matters it is not needed. It would lead to far to discuss it here. In any case one has to grasp the other points of the theory first.

"Various objections have been made earlier by T. Skolem [1926]."

These objections are based on misunderstandings. From the beginning T. Skolem has held that the distinction between sets and classes is a mere verbal artifice even though, as it was explained above, the two concepts are essentially different. Not all classes are necessarily sets. For this reason he understood nothing that followed. The existence of sets is absolute not conditional; hence they are unique. Skolem remarks: "In paragraph 17" (referring to Finsler [1926b]) "the principle of choice" (for circle free sets!) "is 'proved' by invoking the possibility of introducing something which is free of contradiction. Everything goes along very smoothly." One has to respond that the treasure chest opens more smoothly if one has just the right key than by trying to break it open with a pry bar.

"The argument concerning the set M for which the axiom of choice fails has to be taken as a plausibility argument rather than a proof."

It is reasonable to speak of plausibility as long as one thinks the result could be different: We talk of proof when the result is compelling. In the case of the set M, considered previously, plausibility arguments would suggest that it does not exist. Usually one reasons as follows: The set of all ordinal numbers is inconsistent; the set M is of the same power as this inconsistent set; consequently it too should be inconsistent. Here we have a plausibility argument that leads to a false conclusion. We can assure ourselves that the set M satisfies the axiomatic conditions for set theory and hence its existence is secured.

Furthermore, we can prove in the same manner the following remarkable result: *Among the ordinals which can be defined in set theory there is a largest.* Consider the class of all ordinal numbers which have a successor. This is a set because were it to exist it would not contain itself nor be "essential in itself" (in the present instance this is equivalent to saying that it is not contained in any ordinal number). The relation β is well-defined for this set. It contains exactly the ordinals smaller than itself. Consequently it is the largest ordinal number. Attempts to construct larger ordinals must fail for the same reason that the system of all sets cannot be enlarged. One cannot, for example, take out some smaller number and place it at the end to make a largest; this would contradict the definition of the ordinals and that of the sets.

"In the usual system of set theory the formation of the set M considered above is excluded."

This would be the case if one were to insist that the union set axiom is satisfied. The example of the set M shows that this axiom is false in the domain of all sets. There are even sets of only two elements which do not have a union in the domain of all sets. Take for one of these elements the largest ordinal and as the other the set whose sole element is the largest ordinal. The union set would contain exactly the ordinal numbers, and that is impossible.

In any event, the question arising here is whether the axioms of set theory have the task of describing the properties of sets or of discarding all the sets that do not satisfy certain arbitrary requirements.

In the familiar axiom systems, it seems that the second interpretation has been adopted. But it is necessary to remark that one thus obtains only a partial set theory which does not provide a satisfactory foundation. When considering the foundations of analysis one does not postulate axioms that are only partially satisfiable and then ban those parts of analysis that fail to satisfy these axioms. It may be useful to establish some limitations for some particular purposes, but first of all, it is necessary to determine the extent of the entire domain.

Having formed the complete theory of sets one can separate out a partial system which is well defined, general, satisfies the usual axioms, and is useful for applications. The system of non-circular sets has precisely these properties.

Finally I would like to respond to a question raised during the conference (see Gonseth (ed.) [1941, 156]) by T. Skolem: "What does *'non formal thinking'* mean?"

I am entirely in agreement with T. Skolem in thinking that there is no essential difference between formalism and natural language; the language I am using here could be regarded as a kind of formalism. But it is a fact that in mathematics there are things which cannot be explained in a completely fixed language. One has to accept the facts. Take for example the countable ordinals which are undefinable in a given language. There exist such numbers and one of them must be the least. It is well-defined even though it cannot be represented in this language. One has no right however to exclude it from the countable ordinals. The situation can be illustrated by the following example (Finsler [1925]).

On the blackboard one writes the numbers 1,2,3 and the expression "the smallest natural number which is not written on this blackboard." What number is determined by the expression on the blackboard? The answer is: None! Because if there were one, it would be simultaneously written and not written on the blackboard. However, if one only speaks these words without writting them, or if one comprehends them in thought, then these words do represent a definite number: 4.

Just the same, the ordinal number considered above was not representable in language but in thought: It exists just as the number 4 does. This manner of thinking does not lead to eternal silence, since one can draw conclusions the results of which can be expressed in the language. The introduction of imaginary numbers into analysis produced such doubts; but one learned to calculate with them and after a "voyage through the imaginary" one often arrived at important and real results. One learns to operate with things the entirety of which can be represented in language, but which cannot be represented linguistically one by one. After a "voyage through silence" one arrives at important results. A mathematician who is shown the path is surely capable of traveling forward on his own.

152

Lecture at the Mathematical Society in Basel, June 29, 1953. First published as: "Die Unendlichkeit der Zahlenreihe", *Elemente der Mathematik* 9 (1954), 29–35 (MR **15**, 670).

The Infinity of the Number Line

I am happy to be able to speak to your circle here in Basel, with its time honored mathematical tradition, about a time honored theme, namely the natural numbers.

The question whether something infinite exists is also a very old one. Many have said yes, others no; the disagreement continues to this day.

Attempts to prove that the sequence of numbers is infinite, that is, to show that for each number there always exists a still greater one, were made during the last century; I want to mention here:

Bolzano   1851: *Die Paradoxien des Unendlichen* (§13).
Frege   1884: *Die Grundlagen der Arithmetik* (§78).
Dedekind   1887: *Was sind uns was sollen die Zahlen?* (§5).

The later development of set theory showed, however, that these proofs are not sufficient.

Dedekind gave a rigorous development of arithmetic in the natural numbers under the assumption that there are infinitely many things. His reasons for this assumption, however, cannot be considered sound. Dedekind reasoned roughly as follows: Consider the world of conceivable things. The ego belongs to this world and corresponding to each thing there is an idea of that thing. One thus obtains the idea of the ego, then the idea of the idea of the ego, etc., and apparently an infinite series of ideas. But surely, this is only apparent. If one really does attempt to construct this series, then one sees very soon that these ideas can no longer be distinguished from one another, especially if one still does not yet have natural numbers with which to enumerate them. Soon these ideas can no longer be formed any further; the sequence of these ideas stops.

This consideration also shows, however, that it is not self-evident that the sequence of natural numbers is infinite. There could exist a place beyond which one simply cannot advance. If there really does not exist anything which is infinite, then the natural numbers are not infinite either.

Now if we want to settle this question as to whether the sequence of natural numbers is finite or infinite, then we must first say what the natural numbers are, or what we understand by them, otherwise the question has no meaning.

What are the natural numbers in mathematics?

They are in any case not the words "one", "two", "three", "four", etc.: Nor are they the numerals 1, 2, 3, 4, etc. They are not even strokes placed one after the other //// ..., although all these things can very well be used for counting. When we count, we do use the number words "one", "two", "three", "four", etc.; but we cannot proceed very far with these, certainly not into the infinite.

What therefore are numbers, as they occur in mathematics? They surely should be independent of the language which we speak, and above all they ought to be enduring as well, whereas these dashes will certainly pass away very quickly. The numbers which Euler investigated are exactly the same as those which we still investigate today; they really do endure.

The names of numbers and the numerals are nothing more than names or notations for the numbers, in exactly the same way as the word "house" is only a name for a house in which people really live.

If, however, the numbers are enduring, then it follows that they are not of a substantial, material nature; for material things can always be assumed to be transient. Numbers are, therefore, ideal things. This is the first and most fundamental thing which we really do need in pure mathematics: We need ideal things with which we can operate, and concerning which we can make various statements.

In addition to this, however, we still need at least one relation between these things; for we cannot do very much with things alone and no relation between them.

A simple relation of this kind is given with the natural numbers, namely, the relationship 4 bears to 3. We say: 4 is the successor of 3, or, 3 is the predecesor of 4. It is a simple, asymmetrical relation between 4 and 3 which can be represented by means of an arrow: $4 \rightarrow 3$. I have deliberately made the arrow go from 4 to 3 and not in the reverse direction. In the reverse direction we do not yet know how far we can proceed; in the direction used here, we arrive eventually at 1: $3 \rightarrow 2 \rightarrow 1$.

This is a primitive relation between these numbers, a *fundamental relation* which does not presuppose that other relations are given. We can also denote this relation by a letter, say β, and then write: $4 \beta 3$, $3 \beta 2$, $2 \beta 1$.

It is useful to put $1 \beta 0$, and so introduce a zero, which is the predecessor of 1 but which is not a natural number itself. Further, this zero shall not have any predecessor. Each natural number then has exactly one predecessor: Zero has none.

After this preparation we can now define zero and the natural numbers completely.

Definition. *Zero and the natural numbers are ideal things, which are connected to one another by a fundamental relation β and which are determined solely by means of this fundamental relation, such that*:

(1) *0 β x does not hold: Zero does not have a predecessor.*
(2) *If n β 0 or n β m with N(m), then N(n); provided that there are no a and b such that n β a and a β b.*

Here $N(n)$ stands for: n is a natural number. Thus (2) states that n is a natural number if it has exactly one predecessor which is either zero or a natural number. Further, only in this event shall it be a natural number, which means, more precisely, we shall still have to require that:

(3)   *N(n) holds only when necessary.*

This means that n is a natural number only when it follows from (1) and (2) that it must be a natural number.[1]
If $x β x$ were to hold for some thing x, so that x would be its own predecessor, then x could be a natural number. For, if one assumes that x is not a natural number, then (2) will not affect this assumption. Thus by (3), x is not a natural number. From the opposite assumption, that $N(x)$ holds, (2) would again give that $N(x)$ holds. But this assumption is not necessary and so $N(x)$ does not hold.
In contrast to this, if we assume 1 β 0, then 1 is necessarily a natural number; similarly with 2, where 2 β 1, etc. The system of natural numbers is therefore non-empty, it contains the numbers 1, 2, 3, 4. It is uniquely determined; for, given any n whatever, either n necessarily must be a natural number by (1) and (2), or this is not necessarily so, in which case it is not so.
Finally, the natural numbers are also consistently defined by these means, for, nothing impossible is required: If a thing has certain properties, only then is it a natural number, and not otherwise.

---

[1]   *Editor's Note*: (3) is equivalent to the following statement: $N(n)$ if and only if the assumption not-$N(n)$ contradicts (1) and (2).

The Infinity of the Number Line

Thus, in particular, it is not required that for every number m there must exist a successor n with n β m. Whether there is one or not is something which still has to be investigated.

One can now derive propositions about the natural numbers, however, and in particular the step from n to $n + 1$ can be taken. That is, the principle of complete induction can be established. This principle must be so formulated that the existence of $n + 1$ is not presupposed. It can be formulated as follows.

Complete Induction. *Let* A(n) *be any proposition about the natural numbers. If* A(1) *holds, and if from the assumption that* A(m) *holds it follows that* A(n) *holds also, where n is a natural number following m, then* A(n) *holds for all natural numbers n.*

It is definitely not being postulated that to every m there must exist such an n.

The proof goes as follows: Consider the class of all numbers n for which A(n) is true. The number 1 belongs to this class. If m belongs to it and n is a natural number which follows m, then n belongs to it as well; thus all those things n which according to (2) are necessarily natural numbers belong to this class also. But according to (3), these are all the natural numbers. Therefore, A(n) holds for all natural numbers n.

We return now to the question as to whether the sequence of natural numbers is infinite, that is to say as to whether for each of these numbers m there is a number n such that n β m holds.

It may well appear as though this would be openly self-evident in the world of ideal things. There may appear to be no grounds which might make one hesitate to accept that for each number m, there is another n that follows it.

In reality, however, there are such grounds, and that this is so can be seen by considering the ordinal numbers.

First, I will present the natural numbers intuitively. This can be done as follows: Begin with the number 1; then place a new number after each number which we have obtained in this way. We thus obtain the numbers 1, 2, 3, 4, 5, ..., and this apparently continues indefinitely.

Returning to the ordinal numbers: It is natural to start with zero, then comes 1, and then one places a new number after each sequence of numbers which is obtained in this way.

Initially one obtains the finite ordinal numbers 0, 1, 2, 3, 4, 5, ..., which are usually identified with the natural numbers.

After this sequence of finite ordinal numbers, one places a new number ω, then ω + 1, ω + 2, etc., after all these numbers ω + ω then ω + ω + 1, etc. This construction too can apparently be continued indefinitely.

In reality, however, this is not so: One cannot place a new ordinal after the sequence of all ordinal numbers, since this would certainly constitute a contradiction. If one already has all ordinal numbers then there does not exist a new one beyond these.

It can even be shown that there does exist a largest ordinal (cf. Finsler [1941b]), a definite last ordinal number Ω. One can no longer place more ordinals after Ω. There are no more.

Thus, it is not at all self-evident that after each natural number a further one can be placed: With the ordinal numbers this cannot be done. The question remains as to whether the natural numbers continue or whether there is a largest one.

In order to settle this, one must first clarify the reason why, in the case of the ordinal numbers, one finally reaches a stage beyond which one simply cannot continue any further. The reason does not simply lie in the word "all", which is a clear and logically unobjectionable concept. On the contrary, the intrinsic reason lies in a non-satisfiable circle: The rule of construction for the ordinal numbers is of circular nature. It refers quite explicitly to itself, and this circle is ultimately unsatisfiable. The rule says that one is to place a new number after each sequence which is obtained by means of just this very rule, which is only defined here.

And how is it with the natural numbers? The rule for constructing them is circular in precisely the same way; it is also stated that one is to place a new number after each number obtained by means of just this very rule which is defined here. It could become impossible for natural numbers at some stage too. After all, in our earlier definition of the natural numbers $N(m)$ occurs within the definition of $N(n)$. This circle cannot simply be ignored.

If this circle cannot be eliminated, then it has to be shown harmless in our definition of the natural numbers. But this is not so simple: We must first clarify when such a circle might be harmful. For this purpose it is necessary to investigate a system of things which is more general than the natural numbers, a system in which circularity causes actual harm. The ordinal numbers could be considered. But there is a still more general system: The system of pure sets whose elements are again pure sets, is better.

I will now define more exactly what we want to understand by a pure set, or briefly by a *set*. As has been said, it is a generalization of the natural numbers. The difference is essentially this: A natural number always has only one predecessor, whereas a set can have arbitrarily many predecessors. Even when a, b, c, ..., are all distinct the relations $M \beta a$, $M \beta b$, $M \beta c$, ..., can all hold for a set M, where *a*, *b*, *c*, ..., are also sets. One then calls the "predecessors" the *elements* of *M* and writes *M* = {*a*, *b*, *c*, ...}. This also serves to define the term "set"; in reality, however, sets too are ideal things which are given by their relationship to their elements through the β-relation.

The exact definition of a set now runs as follows.

Definition. *Sets are ideal things which are connected with one another by means of a fundamental relation* β, *and which are determined solely by means of this fundamental relation. The following are to hold.*

(a) *Each set determines its elements, i. e. the sets to which it possesses the relation* β.

Thus, if a set is given, then so are its elements. The converse does not hold however. If definite sets are given, then there does not need to exist a set which possesses exactly these sets as its elements. One cannot simply require its existence. A further condition refers to the identity of sets:

(b) *The sets M and N are identical whenever possible.*

This means that whenever the assumption that the sets *M* and *N* are identical does not contain a contradiction, *M* = *N* shall hold.

Thus for the sets *I* = {*I*} and *K* = {*K*} it follows that *I* = *K*. One could say that this is self-evident, because the sets *I* and *K* are not really distinct. A contradiction would result from treating them as distinct. There are, however, cases in which it is not so easy to decide which sets are identical; therefore (b) is required.

There is still one further condition which is necessary, namely:

(c) *M is a set whenever possible.*

Without such a condition sets might not exist. Thus, whenever the assumption that *M* is a set does not produce a contradiction, then *M* shall be a set.

It now follows that sets do exist. For example, zero is the empty set, a set that does not possess any elements; the natural number 1 is the set which contains zero as an element, 2 contains 1 as an element, etc. In this way the natural numbers are definite sets.

The system of all sets is therefore non-empty; it is consistently defined and unique. Nothing impossible is required: only *if* a thing has certain properties is it a set.

There do, however, exist cases in which given elements do not form a set. For example, if one considers all sets which do not contain themselves, i. e. all sets for which $N \beta N$ fails, then there does not exist a set which contains exactly these sets N. It would have to contain itself, if it did not contain itself: and it could not contain itself, if it did contain itself. The reason why this set does not exist is that, once again, a non-satisfiable circle appears in its definition.

It can happen that a set having a circular definition does indeed exist: the set of all sets contains itself and all other sets. No contradiction arises; this set exists and is circular.

There are also sets which are defined in a circle-free manner, the empty set 0, the sets 1 and 2, for example. No circle whatever occurs in their definitions: they are explicit.

The problem of how to distinguish these two cases from each other now arises. Which sets are circle-free and which sets are circular?

Another difficulty appears here: this distinction, between circle-free and circular sets, cannot be given in an explicit, circle-free way. Were this possible, then the "set of all circle-free sets" could also be defined explicitly, and thus in a circle-free way; it would then be a circle-free set. As such it would have to contain itself, and a set which contains itself certainly cannot be characterized as circle-free.

This distinction, between circle-free and circular sets, can therefore be given only by means of an implicit definition; one cannot continue solely with explicit constructions. In order not to be tied to the intuitive concept we will write "c-free" in place of "circle-free" and make the following definition:

(I)   *A set is c-free whenever its elements are c-free and it does not itself depend upon the concept "c-free".*

(II)   *A set is c-free only whenever this is necessary.*

A set is independent of the concept "c-free" if, on the basis of its definition, it stays the same regardless of which sets are classified as c-free. In the first place, one makes an assumption that certain sets shall be c-free and the rest not. As long as it has not yet been established which sets are c-free, such an assumption is permissible. Whenever a set M is uniquely determined by its definition independent of this arbitrary assumption, then that set is independant of the concept "c-free".

The "set of all c-free sets" is dependent upon the concept "c-free"; for, it changes whenever one denotes new sets as c-free. If no set is called c-free, then it would be the empty set. Should all sets be called c-free, then it would be the set of all sets. Furthermore, this set cannot be defined in any other way, without making use of the concept "c-free"; otherwise it would be c-free and would have to contain itself. We will soon show that this cannot be. By contrast, the empty set is independent of the concept "c-free". One can define it as "the set without elements", without recourse to the concept "c-free". This set will always remain the same.

It may be objected that there are sets which depend upon a certain concept and which therefore cannot be defined without making use of this concept. In fact, such examples have already been considered here: the "set of all sets" is dependent upon the concept "all". That is to say, if one could define this set without employing the concept "all", then there would be no contradiction if one were to form still more sets. But this surely cannot happen.

I will now show that a c-free set cannot contain itself. This is derived from postulate (II) as follows: suppose M contains itself, $M \beta M$. Consider the possibility that M is not c-free. Certainly it has at least one element which is not c-free. Therefore by (I) it need not be c-free. Then by (II) it is not. Therefore: *every set which contains itself is not c-free.*

The natural numbers are c-free. This follows from the principle of induction: 0 and 1 are c-free sets. If $n = \{m\}$ and m is c-free, then n too is c-free; for, n contains only one c-free element and is independent of the concept "c-free". Thus all the natural numbers are c-free.

It follows immediately from postulates (2) and (3) be that each number is different from all its predecessors.

Now it is still to be proven that there do exist infinitely many numbers: for each number m there does exist a number $n = \{m\}$. Since the natural numbers are c-free sets, it suffices to show more generally that for each c-free set M there exists a set N which contains it as its sole element, $N = \{M\}$.

In order to do this I will first consider an apparently more complicated case which is, however, in actual fact quite simple. It will be shown to prove existence in non-trivial cases. I will show that the set of all c-free sets exists.

Let it be denoted by U. The assumption that U is a c-free set leads to a contradiction; for, in this case it would have to contain itself and this is not possible. The assumption that U is not c-free does not lead to a contradiction. First, U is really a new set distinct from all the given sets, since they are all c-free. Second, U is dependent upon the concept "c-free"; for, otherwise the subsequent classification of U as c-free could not lead to a contradiction. Hence the assumption that U is a non c-free set contains no contradiction. Thus it is satisfied, and so U exists.

In exactly the same way it now follows, but with greater generality that:

If the assumption that a class U' of c-free sets forms a c-free set leads to a contradiction then U' forms a non c-free set.

For the proof one can replace U with the set determined by U' in the argument above.

We now apply this proposition to the class which consists of one c-free set M:

If the assumption that this class forms a c-free set were to lead to a contradiction, then it would form a non c-free set. It does not, however, for the set $N = \{M\}$ is c-free. After all, it contains only c-free sets and is itself independent of the concept "c-free". Hence the assumption that N is a c-free set cannot contain a contradiction. From this it follows that N exists. This proves that to every natural number there exists a successor, and therefore the sequence of numbers is infinite.

To picture the situation, think of the infinite as locked up. If we want to obtain it, then we have to unlock it. In order to do this, we need a key, and this key must be turned. This turning is circular in nature. If no satisfiable circle is allowed, we cannot obtain the infinite: should it be allowed, the infinite is obtained.

First published as: "Über die Grundlegung der Mengenlehre. Zweiter Teil, Verteidigung", *Commentarii Mathematici Helvetici* **38** (1964), 172–218 (MR **32**, 1126).

On the Foundations of Set Theory

Part II. Defense

[*Editor's Note*: The following is an outline of this rather long paper.]

Preliminary Remarks.

Chapter 3: A Review.
Discussion of Skolem [1926].

Preliminary Remarks

§1. The first part of this construction of the foundations of set theory had the subtitle *Sets and their Axioms*. It appeared in 1926 in the *Mathematische Zeitschrift* [1926b]. The second part was to have treated the number systems. Since the first part met with such a lack of understanding, it seemed preferable to let these investigations, concerning the natural numbers, the continuum and the transfinite ordinal numbers, appear as separate publications ([1933], [1941b], [1954]).

External circumstances delayed an adequate defence of part I for a long time. In fact, I had hoped that the case would finally succeed on its own, at least among those who really, earnestly troubled themselves about the foundations of mathematics.

But now, even quite recently, old, erroneous objections have been brought forward. I therefore find it necessary to go into these objections more fully in order to refute them. In doing this I shall concentrate first on the attacks and reserve a treatment of the further developments for a later occasion.

§2. It is clear that it will not simply suffice to say what is right, but it is also necessary to expose and refute the many false views and errors. It is not always pleasant, for personal reasons, to have to do this but it certainly has to be done for the sake of the case.

In particular, it is necessary that the distinction between true and false be made *in an objective way*. This means that set theory investigates what *is* true, and not only what is *hypothetically taken to be true* (for instance in axioms).

A *formal* "conception of truth" cannot suffice here: It is far too narrow and can neither solve the paradoxes, secure the infinite, nor govern the higher cardinal numbers. A formalist standpoint which admits only finitely long formulas, has only countably many ways of representing sets and therefore cannot exhaust all the possibilities of a non-denumerable system.

Once one overcomes the antinomies, however, there are *no grounds whatever* for rejecting *absolute truth* in pure mathematics.

In actual fact we are concerned here with the *defence and protection of classical mathematics* which enjoyed uncontested validity until the turn of the last century, that is, until the discovery of the set-theoretic antinomies. Thus we are also concerned with *overcoming* the so-called "crisis" in mathematics.

Chapter 1. Point of Departure and Ultimate Aim

§3. I thought that I had made the standpoint of "On the Foundations of Set Theory" sufficiently clear in the introduction to part I [1926b]. It is the standpoint of *classical mathematics*, which has been secured everywhere but in set theory. Since this was misunderstood by many, further comments are necessary.

The numbers in classical mathematics, in particular the natural mumbers but also the real and complex numbers, are not objects that are arbitrarily created by man (how could man ever create infinitely many objects?) but exist quite independently of him. He can only investigate them and undertake research. This was also the view of Frege [1884, §96]: "The mathematician cannot create something arbitrarily, any more than the geographer can: He too can only discover what there is and name it."

§4. An important task, which is often overlooked, is that of deciding whether there are infinitely many numbers or not. This is a question of an objective nature and cannot be solved by mere assumptions.

The system of natural numbers appears to us at first, as far as we can judge it, to be something very simple: After each natural number another one follows. But does this *always* hold for the *whole* number system?

Some effort is required to see that it might be otherwise, that there might be a last natural number which cannot be exceeded. This matter has to be investigated. We can really survey only a tiny part of the number system.

Euclidean geometry first appeared to be something simple, and therefore for a long time it was thought to be the only possible geometry. It was not easy to see that it might be otherwise, namely that lines could only possess finite length. Of course this possibility does actually exist. This had to be investigated too.

If, as many believe, the infinite does not exist, then the number system would be finite. It would not have to turn back upon itself (as a line in elliptic geometry does); it is quite conceivable that there is a last and greatest natural number.

Someone disputing this overlooks the fact that it really is the case with the ordinal numbers.

Every ordinal number can, in a well known way, be made equal to the set of all smaller ordinal numbers. The smallest ordinal number is zero, which is identical to the empty set. Then 1 follows as the set which contains 0 as its sole element, then the ordinal number 2 which is the set which contains 0 and 1, and so on.

Proceeding in this way we obtain the ordinal numbers, each one equal to the set of its predecessors. One ordinal number is greater than another if and only if it contains the other as an element. No ordinal number contains itself.

One cannot form the set of all ordinal numbers, since its definition contains an inherent contradiction. If it were not an ordinal number, then it would still contain exactly all preceeding ordinal numbers, and therefore it would have to contain itself as an element, which is impossible.

In contrast to this, there does exist a set of all those ordinal numbers for which there exists a greater ordinal number. This is the largest ordinal number; it is no longer possible to find a successor to this one. Because of this it is impossible to deduce from its definition that it ought to belong to its own elements. Thus, comparing it to its elements, we see that it is a *new* set and consequently its existence is not thrown into question.

The *class* of all ordinal numbers is uniquely defined, but no set corresponds to it. On the other hand, the *union-set* of this class does exist: It is once again the *greatest ordinal number*.

This has nothing whatsoever to do with the infinite, about which nothing at all has been said here. The greatest ordinal number could be finite, and would then coincide with the greatest natural number. In this event the number system would be finite. That this is not the case is therefore in no way self-evident.

Actually the supposition that every natural number must have a successor could indeed contain a contradiction, because the definition of an *arbitrary* natural number, like the definition of an *arbitrary* ordinal number, contains a circle (see [1933], [1954]). With the ordinal numbers this circle finally prevents further successors; it must be shown that the natural numbers do not behave in the same way.

In the sequence of natural numbers there is a definite first number, the number 1. Why should there not also be a definite last number? If this is not the case, should we not be able to prove it? Must there be only a mystical belief in the infinite? Belief in an absolute mathematical truth is well justified, as soon as the antinomies have been overcome. But the distinction between true and false in mathematics ought not be a matter of faith: It must be investigated.

That the good God has made the natural numbers will be readily conceded. He has, however, left it to us to find out how many there are. The good God has also made the atoms in our world but it is highly probable that there are only finitely many.

§5. The question as to whether there do or do not exist infinitely many prime numbers is very well known. It is usually accepted that there do. This was proved by Euclid. Euclid's proof is valid only if one already knows that there are infinitely many natural numbers. But is this known? No, most people do not know it. On the contrary, it is only assumed. But then, one does not know whether there are infinitely many prime numbers. Euclid's proof would be in vain.

An axiom of infinity is therefore used and it is said that the proposition is correct. But if the axiom is not true, then the proposition is false. Thus, once again, nothing is known.

The axiom of infinity states that there exist infinitely many things. In the perceived world it is very probably not fulfilled. So an ideal world is needed. Do there exist infinitely many things in this ideal world? If this cannot be shown, then the mathematics of the infinite breaks down. One would have only hypothetical statements and one could not even know whether Euclid is right with his assertion.

Some say that infinitely many things are not needed in mathematics; all one needs is that whenever arbitrarily many things are given, there will always exist another. But this leads no further: It is exactly the same, only expressed differently.

§6. It is by far the most important task of a real building of the foundations of mathematics to clear up this point and show that there are infinitely many numbers. This is not easy, but also not impossible (see [1933], [1954]). If it is contended that an "absolute proof of consistency" for Peano's axiom system for the natural numbers is "not possible" or "not conceivable", then it should be remembered that once upon a time flying was declared to be impossible and that the brothers who, in spite of this, did accomplish it were called liars.

§7. Of course this cannot be done with purely formal methods. A system of formulas which "can be written down" is necessarily only finite, and thus compatible with the hypothesis that the universe is finite. But if the universe is finite, then it cannot be shown that it contains infinitely many things or infinitely many "symbols"; in this event it would simply be false. There would be no additional things that could be found.

The infinity of the number system can never be calculated, but it can be seen nevertheless. But, one must try hard enough.

It is disgraceful and fraudulent to simply assert that there are infinitely many numbers without knowing whether this is the case. If a child asks: "Is it true that to every number there always exist a greater one?" Can one then *in good conscience* answer: "Yes, that is true."

It is also *unworthy* to pretend in mathematics that there are infinitely many things, especially when one does not oneself believe in this. Many deny the infinite and nevertheless want to teach differential and integral calculus and operate with convergent series and are of the opinion that one can always place another symbol after an arbitrarily long sequence of symbols. But they cannot guarantee that every sequence can be extended.

It is *untenable* to claim that it follows from basic intuition that every natural number has a successor. With equal justification one could assert the same for the ordinal numbers and that is false, as has been shown above.

If, however, one finally says that it suffices to show by using formal methods that a contradiction cannot be derived from the axiom of identity, then this is just as degrading as when a criminal says that he permits himself to venture anything, just as long as he is certain that he will not be caught, or at least not be sentenced under the law. Whoever has committed a murder is a murderer, even if it cannot be proved against him.

§8. It is an error to consider a system of formulas as being something "more exact" than the numbers and their relationships.

First of all, a formula consists of "symbols". What is a symbol? Perhaps it is a question of the materially visible signs that can serve for momentary communication. But if nothing else exists, what are these signs communicating? If symbols are used for the communication of logical concepts or relationships, then the logical concepts or relationships form the foundation, not the symbols. A mere game with meaningless symbols remains meaningless.

Moreover, the symbols are inexact and transient. They can never serve as foundations for an exact and permanent mathematics.

What Euclid did in attempting to define the point and the line by means of intuitive properties, would today be rejected in a study of the foundations of geometry: What matters is only the relationship which can arise between point and line and not "what they look like".

To define numbers by means of symbols is to take a step back; here too it is only the relationship between the numbers which matters and not their "appearance".

If one did in some way manage to define the "symbolism" exactly, even then the formulas constructed from them could not be a substitute for thinking. Formulas can be useful for many purposes, but a thinking formula is no more real than a thinking machine. The formulas only represent certain externally given relationships between things; the essential, intrinsic interconnections are often neglected, and for this reason serious mistakes can arise (see [1944], and §13 below). For instance a formally consistent system can be entirely wrong in content (see [1926a]).

Numbers, as ideal things, are not visible in the same way as symbols which have been written down, or objects of daily life; but they are no less clearly recognizable for one who seriously concerns himself with them. This does not mean that every single number taken on its own shall always be sharply identifiable; but one can indeed recognize that the numbers form exactly defined systems which are invariant and permanent. No one has ever seen the exact number π; in spite of this it is exactly, clearly, and uniquely defined. The same holds also for pure sets which are generalizations of the natural numbers.

§9.　Pure mathematics, which we have to investigate, is independent of our human shortcomings. Whether a mathematical proposition is true or false has nothing whatsoever to do with whether we can decide this or not. The law of the excluded middle is the basic requirement for mathematics. That there exists such a mathematics, and indeed one which is supremely rich in content, is shown by experience and by logical thinking. To demonstrate this in detail is our task when we study the foundations of mathematics.

Mathematics ought not to be unnecessarily or arbitrarily restricted by admitting only those things that are constructible in a certain way, for example. The domain of the real numbers goes beyond this and we cannot permit the prohibition of some real numbers. Everything which satisfies the law of the excluded middle, that is everything unambiguous and consistent, is to be included in pure mathematics.

§10. In his *Grundlagen der Geometrie* Hilbert [1913] showed that Euclidean geometry is consistent, provided that the arithmetic of the real numbers is consistent. In *Grundlagen der Mathematik* it must be shown that the whole of mathematics is consistent without further preconditions. In particular, the classical arithmetic of the real numbers, on which geometry is based, is also consistent. It is impossible to show this by formal means alone. Therefore it is really better not to regard formalism as a foundation of mathematics. Mathematics is not merely a formalism.

Similarly one ought not to speak of a "proof of consistency for number theory" if the natural numbers are *assumed* to be already given. In the foundations of geometry, it is not presupposed that geometry is already given. One investigates whether the geometrical axioms are complete and compatible with one another.

In order to obtain purity in laying the foundations there must be no misleading designations.

Furthermore, investigations of *results* used in mathematics, arithmetic in particular, should not concern themselves with whether they are "more or less secure" but only whether they are *true* or *false*. If one speaks of results that are "more secure", then, in doing so, one admits that they are *insecure* (which is understandable when they are partly false). What good are mathematical proofs if they give *no certainty*? Certainty is not attained through fixed regulations and prohibitions, but through the revealing and eliminating of mistakes which have been made.

We are not dealing with some philosophy here, but with pure mathematics and with logic.

§11. Logic, that is correct reasoning, must be employed if one wants to establish mathematics. Logic can be investigated in detail, but it is to be considered as being intrinsically fixed and determined and also as being independent of us and our thinking. Thinking must align itself with logic in order to be "right thinking". Logic is just as invariant and permanent as are numbers and pure mathematics; only our knowledge of these things can change.

Thus, what is correct according to logical thinking or logical reasoning does not depend upon conventions, but is determined in an absolute way; there is an absolute distinction between true and false. In this sense there can only be *one* logic. There is also no sense in wanting to prove the consistency of this logic. It is consistent;

otherwise it would not be logic. Contradictions arise only through mistakes, that is through disregard of logic.

Logical thinking is not identical to "logical calculation", "formal logic", or "logistics". Such things can at most represent only a part of logical thinking, but in each case it must be checked whether it really does satisfy the requirements of logic. These things themselves must be "right", and indeed not only in a formal sense but also in meaning. What was said above (under §8) about formulas holds here as well.

If one believes that mathematics or logic can be replaced by formalism, then one overlooks the fact that the cardinal error which leads to the antinomies consists in just this fact, that one restricts oneself solely to the formal interrelations between the objects under consideration and does not observe the real, inner relationships.

Thinking can be supported by formalism, but it cannot be replaced by it. A sentence which is grammatically correct in construction can be quite false in content, and a formula produced in accordance with apparently sound rules also need not always express true facts.

There are, however, very many things which are, by their very nature, not representable by formulas. This follows from nothing more than the denumerability of the formulas and the non-denumerability of the things of analysis for instance. Thus the whole of mathematics can never be obtained by formal methods alone. From the absolute point of view these give a mere fragment.

§12. Anyone who can think logically ought to know what a contradiction is. The "principle of contradiction" belongs among the basic statements of logic. It can be said that a contradiction arises as soon as something must be simultaneously both true and false. Here "false" means the same as "not true". Thus, if one contradicts oneself or, what amounts to the same thing in pure mathematics, if one contradicts some already established fact, then one has a contradiction.

A contradictory assertion is never true: And an inconsistent thing, a thing whose properties are self-contradictory, can never exist. On the other hand, consistent things can always be taken to exist. In pure mathematics, existence means nothing more than freedom from contradiction.

It is obvious that if mathematics is to be correct, then it cannot contain any insoluble contradictions, and thus no antinomies; nor can antinomies appear if one is careful never to contradict oneself. Where no contradiction is put in, none can come out. The antinomies of set theory can be clarified and dissolved in this way. If one retains the antinomies, however, or only tries to detour around them, then one cannot distinguish between true and false. In that case, any nonsense could be asserted, because from a single contradiction all others follow.

It is, for example, a contradiction to assert that the existent, that is consistent, sets cannot all exist at the same time. This would mean that existing sets do not exist, and that is certainly *blatant nonsense*!

Even distinguished mathematicians have been led astray into false opinions through unclarified antinomies. But we do not have to persist in error!

§13. If I assert "I lie", or more explicitly "that which I am now asserting is false", then I contradict myself and thus maintain something false. That is to say, every assertion, just by its being an assertion, implicitly contains the allegation that it is true. If, however, at the same time it explicitly asserts that it is false, then this constitutes a contradiction. It amounts to a simultaneous assertion that something true is false or something false is true: The assertion as a whole is therefore false. One cannot claim in that event that the assertion is true; it would be asserted of something true that it is false and that assertion is false (see [1944, §2]).

Why then is this simple explanation still not yet recognized? Apparently because it cannot be represented with the known formalism. Formalism overlooks the implicit assertion and can therefore merely state a contradiction, or at best evade it, but cannot solve it.

But if one cannot solve even this simplest of paradoxes correctly, how then will one fare with the more difficult antinomies of set theory?

When one looks for the mistake in the wrong place one risks making additional errors.

It is remarkable how many different opinions and theories have been developed in connection with the antinomies, instead of simply asking what is true and what is false, and then acknowledging the truth.

The liar paradox is usually dismissed as being *meaningless*. That is very convenient, but *wrong*. A meaningless assertion is not false and a false assertion is not meaningless. Now, if the assertion "that which I am just now asserting is false" were meaningless, then it would be false, because it would be asserted of something meaningless that it is false, and therefore meaningful. There would be an assertion of something that does not hold. Therefore, the quoted assertion cannot be meaningless, otherwise it would be false and consequently not meaningless. Everywhere this argument is *passed over in silence*.

§14. The *antinomies* have called forth, or reawakened, a "horror infiniti" even though they have *absolutely nothing whatever* to do with the infinite.

The liar paradox seems to lead to a contradiction, but the infinite is simply not involved in the matter. In the *whole of classical analysis* not only does the infinite appear but even the uncountable infinity of the continuum, yet *there are no antinomies at all*. In actual fact the antinomies stem solely from *defective circles* and from the belief that one can "construct" things which are defined in a faulty way.

This also holds for the "paradox of finite definability", which does not belong to classical analysis and which can be so restated that even there the infinite plays no part (see [1926a]).

Now the opinion has arisen that everything which is "finite" is evident, admissible and incontestable, while the *actual infinite*, in particular the number system as a whole, must be rejected.

Both are wrong. This view is a *surrender* to the antinomies. One believes that it is necessary to withdraw into the cramped shelter furnished by the finite.

The natural numbers are usually defined by the act of placing symbols one after another. This, it is supposed, can proceed arbitrarily far. Just try to make a trillion strokes and *then still one more*. This is nonsense. Number theory, however, considers far larger numbers. It will not do to regard numbers as symbols. In making the induction from n to $n + 1$, one needs *all* natural numbers up to the one under consideration. Are these all only symbols? What then is a symbol? As one really cannot make arbitrarily many strokes, the natural numbers *must* be considered as being *ideal* things, which cannot all be written out individually but which nonetheless all do *exist* in an ideal sense.

If, however, one admits the natural numbers as *ideal things*, then they are *independent* of us and our capacities and one must then be allowed to speak also of the *totality* of these numbers. The existence of these numbers is *timeless*, and therefore a "growing" sequence has no meaning. We are concerned with a *definite* aggregate. There are really *no grounds at all* for rejecting the sequence of natural numbers *as a whole*; to do so would be a thoroughly *arbitrary prohibition*.

The question as to whether the series of natural numbers is *terminating or non-terminating* is quite *another* matter. Here we would like to know whether for *every* natural number there exists a successor or not. It has already been mentioned that the existence of a *non-terminating* sequence of numbers is *not a self-evident fact*. The definition of an *arbitrary* natural number is *circular*; it refers to the concept of a natural number. That such a circle can stop the sequence has been shown in the example of the ordinal numbers, treated in §4.

Since this same circularity occurs with the natural numbers, each one of which is finite, it follows that the finite is *no less in doubt* than the infinite, which would appear naturally if it were already known that every number has a successor. Thus it really makes no sense to admit the idea of so-called "existence in itself" for arbitrarily large finite systems but to reject it for infinite ones.

It is just as unfounded to maintain that one could verify an arbitrarily large finite number of instances of a proposition of arithmetic but not verify infinitely many. In reality both are equally possible or equally impossible. Here on earth we can individually verify only a few of the possible instances, even with the very best computers; we certainly cannot verify either infinitely many or arbitrarily many. Yet if one allows that proofs be carried out *in abstracto*, through an *ideal* procedure, then all the proofs can very well be carried out *at the same time*: It is quite irrelevant whether there are finitely or infinitely many. That a finite procedure has advantages over an infinite one is of course not disputed.

Pure mathematics, however, ought not be bound by human inadequacies; these do not affect it.

Chapter II. The Axiom System

§15. The first objection to be raised and published against *On the Foundations of Set Theory, Part I* [1926b] originates in Reinhold Baer [1928a] and concerns the axiom of completeness in this set theory. In a subsequent reply I have rejected this objection [1928]; in fact it was already discussed in *On the Foundations of Set Theory, Part I* and refuted there. I did not think that I would have to answer Baer's remark [1928b] beyond what I have already said. Surely everyone who thinks the matter over correctly must see which of the two is right.

Apparently some have not given the matter enough thought, and it has simply been concluded that he who "has the last word" is right. My further observations [1933] were not understood either, nor was any attention paid to them; so the objection was left unexamined and, heedless of the consequences, others adopted it. Even quite recently it has been brought forward against "On the Foundations". For this reason it is necessary to deal with these things in detail. In the course of doing so attention must be paid to the fact that apparently unimportant matters can turn out to be important for full understanding, all the more so if some are of the opinion that Baer has "improved" the statement of my axioms, which is definitely not the case.

§16. Baer gives the first axiom, the "axiom of relation" the following form.

"For arbitrary sets M and N – these are elements of Σ determined by the axiom system –, it is uniquely decided whether or not the ∈-relatiom holds between M and N, that is whether $M \in N$ or $N \in M$ is true." [1928a, 536]

I have *deliberately* and *with good reasons* chosen the inverse relation β in place of the usual ∈-relation as the *initial relation*, so that $M \ \beta \ N$ means: M possesses N as an element.

This is of significance for an understanding of the antinomies, because a set always determines its elements, but it cannot be concluded conversely that given arbitrary elements there is a set which corresponds to them.

It is of grcat importance to know what elements a set has, but quite inessential to know in what sets it is contained. The empty set is characterised by the fact that it possesses no elements. The empty set is not affected by the sets to which it belongs. Do we need to say explicitly of every set that it must not occur as element of the empty set? And incidentally from where would one get all those sets as long as one did not yet have the empty set?

Consider the difference between these formulations: "This box is empty" and "Every single thing is not contained in this box". It seems that modern research into foundations gives preference to the second formulation.

The empty set, as an object which possesses the b-relation to nothing, is very *easy* to define; the existence of other things is not presupposed. If, on the other hand, one requires of every set that it is not allowed to possess the ∈-relation to the empty set, then this is *infinitely complicated*, as soon as there are infinitely many sets. It is thus *emphatically not a matter of indifference* which relation is chosen as the initial relation. It is the β- and not the ∈-relation which is important for the definition of a set.

Besides, the ∈-relation intuitively suggests "being contained in", which (except in quite simple cases) must be avoided in carefully laid foundations, because it can easily lead to mistaken conclusions.

In an orderly laying of foundations one ought to pay attention to such matters.

§17.   I had formulated the axiom of relation as follows: *For arbitrary sets M and N it is always uniquely decided whether M possesses the relation β to N or not.*

Contrary to Baer's version, it is clearly stated from the outset that a relation *directed from M to N* is being dealt with and *not* a *reciprocal* relation *between M and N* as is the case with an identity or equivalence relation.

In Baer's drafting of the axiom $M \in N$ could hold when M is identical to N; this difficulty remains because later Baer does not introduce identity but only an *equality* of sets.

If one already knows what $M \in N$ signifies in set theory, then one knows that one is dealing with a *non-symmetrical* relation. But *from where* does this knowledge come? At the beginning of a theory it is *impossible* to presuppose the very first concept which has to be introduced and explained *as being already known.*

The asymmetrical form of the symbol ∈ does not decide anything. For equivalence relations one usually employs symbols without left-right symmetry, ≈ or ≡, for example. Conversely symmetrical symbols usually represent asymmetrical relations, for example "$a \mid b$" for "a is divisor of b". *One ought not, however, make newly introduced concepts implicitly dependent upon the form of the notation used for them.*

Thus, if one interprets "$M \in N$" as signifying the identity of M and N, and the next axiom of Baer, concerning the equality of sets is added, then as models for the set theory of Baer one could take systems of arbitrarily many non-identical things which are all equal to one another, and concerning which *otherwise nothing is stated.* These are surely *extremely uninteresting* set theories.

If, however, one improves Baer's version of the first axiom and introduces the ∈-relation of $M \in N$ as a relation *from M to N*, then this results in a *new discrepancy* since Baer says later [1928a, 538] that a set A stands "in the ∈-relation, $B \in A$" to certain sets B. This is *wrong*; A does not stand in the relation ∈ but in the inverse β-relation to B.

§18.   The elements of a set M, the elements of these elements, and so on, are the "*sets essential in M*". Baer defines these in such a way that the set M always belongs to them. This is *not allowable* for an investigation of the paradoxical sets and the antinomies. It constitutes a *very important* difference whether a set "is essential in itself", when it contains itself as an element, for instance, or whether this is not the case, as happens with the empty set. The former kind is always circular. According to Baer's definition all sets would be essential in themselves and thus they all would be circular (see Finsler [1926b, §13]; nothing more would be left over to be circle-free.

§19. A system of sets that contains all elements of each of its members was called a "transitive system" in "On the Foundations of Set Theory, Part I" [1926b, §7]. The sets that are essential in M are those sets which belong to every transitive system that contains all elements of M. The system of sets essential in M was denoted by Σ_M.

Two transitive systems of sets Σ and Σ' are then, in general, taken to be *isomorphic* whenever there exists a reversible, single valued and relation-preserving mapping between the sets of Σ and the sets of Σ'. That is, there is a one-to-one mapping for which: If A and a are mapped onto A' and a' respectively with A β a holding, then A' β a' holds as well, and conversely.

Baer on the other hand defines the transitive systems Σ_M and $\Sigma_{M'}$, which according to him also contain the sets M and M' respectively, as being isomorphic *only if* M is mapped specifically onto M'. This is an *unusual* requirement for an isomorphism nor does it make good sense. Under certain circumstances it may not be at all possible to see from the *systems* alone what sets within them are meant by M and M'. *Concepts should not be made dependent upon the notation employed.*

For example, $A = \{A, B\}$, $B = \{A\}$. Here the systems Σ_A and Σ_B are *identical*; they consist of the two sets A and B. According to Baer they are *not isomorphic*, because $M = A$ and $M' = B$.

This example, which by the way can be found in "On the Foundations of Set Theory, Part I" [1926b, §7] shows that an isomorphism of the *systems* Σ_M and $\Sigma_{M'}$ taken in the usual sense, *does not suffice* for identifying the sets M and M'. One can of course map A onto A and B onto B; the system composed of the two sets A and B is isomorphic to itself. A cannot, however, be identified with B because A contains itself but B does not.

I have therefore called the *sets* M and M' *isomorphic* whenever the systems Σ_M and $\Sigma_{M'}$, which are essential in M and M' respectively, can be mapped one onto the other in a one-to-one, reversible and relation-preserving fashion such that the elements of M are mapped onto those of M'. Because of this the second axiom, the "axiom of identity", could be briefly formulated thus: *Isomorphic sets are identical.*

§20.   Now, in the original version, as given in the "On the Foundations of Set Theory, Part I" [1926b, §7], I made an *inadmissible simplification* in the *definition of isomorphism* of sets. I believed that it sufficed to add to the systems Σ_M and $\Sigma_{M'}$ the sets M and M' respectively and then to require that M can be mapped onto M', because then the elements of M are mapped onto the elements of M'. It is not easy to see what is wrong with this without producing an example.

Let J be a set which contains itself as its sole element, $J = \{J\}$. Further, let L be a set that does not contain itself but which contains the set J as its only element, J being different from it, so that $L = \{J\}$ is satisfied with $L \neq J$. Here $L = \{J\}$ does not mean that L is *that* set which contains J as sole element, but only that L is *one* such set.

The systems Σ_J and Σ_L both consist of only the set J, and according to the appropriate definition of isomorphism the sets J and L are isomorphic. Therefore, according to the axiom of identity, they would have to be identical. This means that no such set $L \neq J$ can exist in a system Σ which satisfies the axioms.

According to the mistaken definition J and L would not be isomorphic, because J contains itself but L does not; there would be two different sets with the same elements in contradiction to proposition 5 of "On the Foundations of Set Theory, Part I" [1926b, §8] which states that two sets which possess the same elements are identical.

G. Köthe brought this discrepancy to my attention. To my knowledge this is the sole mistake to be found in "On the Foundations of Set Theory, Part I". I *corrected it immediately* [1928] by returning to the original definition.

§21.   In Baer's version the second axiom reads as follows:

> $M = M'$ – where M and M' are sets – if and only if Σ_M and $\Sigma_{M'}$ are isomorhic, that is if there exists a one-to-one mapping of the elements of Σ_M onto those of $\Sigma_{M'}$ such that:
> 1. M is mapped onto M';
> 2.   If A_i has image A'_i ($i = 1, 2$) then from $A_1 \in A_2$ it follows that $A'_1 \in A'_2$ and conversely. [1928a, 536f.]

This version is incorrect for the reasons given in §19 and in §20. Moreover, in place of identity only a relation $M = M'$ is defined.

Shortly afterwards a "*weaker axiom of equality*" and a "relation of equality for set theory" are mentioned. This gives the impression that several sets can be *equal to each other* in the system Σ in much the same way as many line segments can be congruent to each other in geometry.

In an *unambiguous* set theory this is obviously *not admissible*.

If, for instance, a set possesses two equal sets as elements, how many elements does it then possess? One could, as is otherwise customary in set theory, require that this cannot occur, since the elements of a set must be "well differentiated". Against this there stands the other requirement that from $a \in M$ and $a = b$ there must always follow $b \in M$. Thus a set would have to contain everything that is equal to one of its elements. *How many of them are there*?

One might have a makeshift "relativity of cardinal numbers" and say that "set theoretically" there is only one element, even though in reality there are many. One would have to speak of "set theoretically uncountable" sets even though they actually contain only countably many elements.

Such obscurity and ambiguity has no place in a *meaningful* set theory. The system Σ of sets is permitted to contain always only *one* specimen of each set and this has to differ *essentially* from all other sets in Σ in terms of the β-relation alone. This is the meaning of the second axiom.

§22. Frequently the proposition which states that sets which possess the same elements are identical, is to *define* identity for sets. With arbitrary sets of sets, however, this is *not admissible*. How should one know whether the elements are the same if one does not already know which sets are identical? How could one decide whether $J = \{J\}$ is identical to $K = \{K\}$ or not (see [1926b, §8])? An example in "On the Foundations of Set Theory, Part I" [1926b, §18] shows that this indefiniteness does not arise only in connection with sets which contain themselves or which are essential in themselves. This definition of identity for sets contains a *vicious circle* and is therefore useless.

Baer appears to be heedless of this. In the works of Fraenkel [1926] and Vieler [1926] which he quotes, this circle is *disregarded*.

A similar objection could be made against a definition of identity in terms of isomorphism: A one-to-one mapping has an exact meaning only after the identity of sets has been decided. Even given this objection, it would still be impossible to have arbitrarily many "indistinguishable" sets enter into the theory. *Different* sets must be distinguished in *essence*, namely by the β-relation.

From the *axiomatic* (not formal!) standpoint which was adopted in "On the Foundations, Part I" the objects of the system Σ are considered, as is customary, as being given entities whose identity is determined in another way, and the "axiom of identity" merely excludes from the system Σ those things that do not satisfy it.

Seen from a *higher standpoint*, however, it can be required that sets be characterized *by the β-relation alone without the use of any other property*. From that point of view the objection described above would be justified.

From this standpoint the systems of sets must be determined not merely as a "structure", that is only up to isomorphism, but *uniquely* (as in Finsler [1954]). The identity of sets is then defined so that two sets given by means of the β-relation are always considered to be identical *whenever possible*. By these standards two sets are always identical if the *assumption* that they are identified produces *no contradiction* between the β-relations involved.

Suppose that the sets A and B were previously given by the relations $A \beta A$, $A \beta B$, $B \beta A$. It would actually follow that $A = B$, because this is compatible with the conditions; it would not be allowed to require that $A \neq B$. The set A would be identical to the J-set, $J = \{J\}$, but it is different from the empty set, because $A \beta A$ holds, whereas the empty set never possesses the relation β to another set. Were we to attempt to identify A with the empty set a contradiction would result.

§23. Next, Baer stated that one is always entitled to speak of a *set theory* whenever one has a system Σ, in which axioms I and II, that is the axioms of relation and identity, are fulfilled.

It is correct, as Baer mentions later, that these axioms do not contain *any condition of set existence*. This is of the *greatest significance*; for, it is just the non-satisfiable existence conditions, those that require the existence of certain sets even when their definitions contain a contradiction, that lead to the antinomies.

In any system which satisfies the first two axioms, the *only* sets that actually exist are those whose definition contains *no contradiction*.

Now, the *system* Σ contains *all* those sets which are free from contradiction, all those which exist, or all those which occur in a system in which isomorphic sets are identified. This total system *exists* because *nothing impossible* is required by its definition. It includes only the *existing* sets. It is nonsense to suppose that existing sets should cease to exist when they are brought together (cf. §12).

The total system Σ satisfies axiom III, the "axiom of completeness". It *cannot be extended* because there are *no* sets *still remaining* after *all* sets have been brought together.

§24. In order to attack the axiom of completeness Baer asserted the following proposition which is, in fact, false.

> Let Σ be a system of sets which satisfies axioms I and II, then either:
>> Σ *is consistent*, that is, there are two sets A and B in Σ such that $A \in B$ and $A \notin B$ hold simultaneously, or else
>> Σ *is capable of extension*, that is, there exists a system Σ^* of sets such that:

> 1. It satisfies I and II;
> 2. If A and B are contained simultaneously in Σ and Σ^* then $A \in B$ holds in Σ^* if and only if it holds in Σ;
> 3. There exists a set which is contained in Σ^* but not in Σ. [1928a, 537]

Thus Baer allowed the possibility that the system Σ is inconsistent. But what is an inconsistent system of sets? A statement or assertion can be inconsistent, but then it is false. A definition can be inconsistent, but in that case it will not be satisfiable. An axiom system can be inconsistent, but then nothing would satisfy it either. But how can a system of sets be inconsistent? Inconsistent systems of sets *do not exist*.

Concerning this Baer writes in the course of his "proof": "Thus $N \in N$ and $N \notin N$ are true simultaneously which means that Σ is inconsistent". In this, N is meant to be a set belonging to Σ. So the system Σ would be a system of sets which, by hypothesis, satisfies axiom I, but which at the same time does not satisfy this same axiom. How can a system simultaneously satisfy an axiom and fail to satisfy it? What has happened to the law of contradiction?

A system Σ can satisfy axiom I or it can fail to satisfy it. To do both at the same time, however, is *impossible*.

One could perhaps suppose that the expression "Σ is inconsistent" means that Σ is *inconsistently defined*. But against it remains the fact that Σ is assumed to be an actually *existing system* of sets.

The way Baer expresses himself shows clearly (and later this becomes still more apparent) that he has not come to terms with the antinomies. He permits the antinomies, and thus allows contradictions. Nothing worse can happen in mathematics. In spite of this he is insistent on criticizing and will not take into consideration the fact that once there is an antinomy anything and everything can be proved or refuted. But such "proofs" are thoroughly worthless!

§25. The first "possibility" of Baer's propositions in §24 thus proves to be *impossible*. The second alternative, that Σ is necessarily "capable of extension", is also *false*, as has already been shown above. If the system Σ does not yet contain all sets, then it can of course, be extended; that is trivial. But if Σ contains *all* sets, i. e., all that there are, then this is no longer possible, for, given all sets, there does not exist still another.

How many "logics" will be needed before it is seen that there *cannot* always exist *still one more*? Once one already has *all*, one cannot have more: This surely ought to be understood by every reasonable man, as long as he has not become so confused by the circumstances surrounding the antinomies that he has forgotten how to *think logically*. A person *cannot be prohibited* from speaking about *all sets*.

§26. In order to "prove" his proposition, Baer considers the system **N** of exactly those sets from Σ which satisfy $A \notin A$. This is admissible. Then he says:

> There now occur two possibilities: Either
> 1. There exists a set N in Σ which corresponds to **N**, so that from $A \, \eta \, \mathbf{N}$ it follows that $A \in N$ and conversely; or
> 2. there does not exist such a set N in Σ. [1928a, 537]

In the first case he concludes that Σ is inconsistent. To be fully correct it would have to be stated that this case cannot occur.

In the second case Baer tries to "adjoin to Σ a set N corresponding in the above sense to **N**". By the hypothesis of this second case no such set N exists in Σ, and if Σ really contains all sets, then this means that there can be no such set whatever. *How can one add to Σ a set which does not exist?*

It is not only that the proof of the existence of the set N is wrong: Such a proof is impossible. The set N is inconsistently defined and therefore cannot exist at all. So the system Σ cannot be extended by adjoining it.

Thus, the proof of Baer is *wrong* and the proposition to be proved is *false*.

§27. Baer does not object to speaking of "*all possible consistent set theories*". He contends, however, that "*their union* does not yield a *consistent set theory*", and that "so to speak" the "upper limit" of all set theories is inconsistent".

The union itself is still taken to be correct, only, in this manner of speaking, the result is not a consistent set theory, but an inconsistent system of sets. It has already been shown in §24, however, that this is *meaningless*; inconsistent systems of sets do not exist.

It is apparent from this, along with the earlier description "Σ is inconsistent", that it is not merely an incorrect mode of expression that is at stake here, but really *defective thinking*.

§28. In connection with the proposition given here in §24 Baer refers to theorem number 10 of Zermelo [1908]. This could easily give the impression that something similar is to be found in Zermelo. That, however, is definitely *not the case*.

Theorem number 10 of Zermelo [1908] states that every set M possesses at least one subset which is not an element of M. This holds for the domain D being considered by Zermelo. From the theorem it follows, according to Zermelo, that not all things of the domain D can be elements of one and the same set, which means that *the domain D is not a set itself*. This is correct.

Zermelo, however, *in no way* concludes that either D is inconsistent or D must again be capable of extension. That is a *big difference*!

Now, in the total system Σ there actually does exist a *set of all sets*, which thus corresponds to the total domain Σ. One can conclude from this that not all of Zermelo's axioms are satisfied in this domain; in particular "the axiom of separation" is not satisfied.

In the system of *circle-free sets* the axioms of Zermelo are satisfied. This system must not contain, as Zermelo shows, either the set of all circle-free sets or the set of all circle-free sets that are not members of themselves. Nevertheless, this system of sets *cannot be extended*. The two sets just named do indeed exist and are moreover identical, because no circle-free set contains itself. This set is a *circular* set which, for this reason, cannot be inserted into the system of circle-free sets without producing a contradiction.

§29. Baer concerns himself next with axiom III, the "axiom of completeness". It has already been observed in §23 that the total system Σ satisfies this axiom. Axioms I, II, and III are, *contrary* to what Baer maintains, *simultaneously satisfiable without contradiction*. It is hardly surprising that this goes against his false statement.

Baer goes on to claim that the union of *all* possible set theories is a false inference. Earlier he would have allowed it. *Why* should it be impossible now? All consistently existing sets (according to Baer those occurring in a "set theory") are simply brought together; in doing so isomorphic sets are taken to be identical. This yields a consistent system of sets, which represents the desired union. This consistent union provides a *set theory which is free from contradiction*.

It is an *error* to *reject* such a union, for there are *no grounds* for such a rejection.

§30. The objection of Baer will be found, as mentioned in §15, in "On the Foundation of Set Theory, Part I" [1926b, §12] and was already refuted there. There was a reference to "Are There Contradictions in Mathematics?" [1925] to which Baer obviously *did not pay any attention*. He could have found a complete refutation of his view on the next to last page of that paper.

In order to attack the refutation given in "On the Foundations of Set Theory, Part I", Baer contends that the object N, which he wants to adjoin to the system Σ, really is something "new", that is, an object not belonging to the total system Σ. Thus it *belongs to none* of the systems satisfying axioms I and II; but on the other hand, it *satisfies* these axioms when joined with Σ. This is an *outright contradiction*!

If one object can be united with another object to form a specific system, then it belongs to at least one such system; this cannot be helped. It cannot at the same time belong to *no* such system.

If the thing N, however, belongs to a system satisfying axioms I and II (here Baer's mode of expression is not quite clear), then it is *not a new* thing, that is, a thing not in the total system Σ!

The contradiction can *only* be solved in the following way: There *does not exist* a thing N with these properties.

So in reality the contradiction lies in the definition of N and *not*, as Baer maintains, in the axioms. If one includes in the total system Σ only the sets which belong to one of the systems satisfying axioms I and II, then axiom III *does not appear* here *at all*!

This shows clearly, contrary to what Baer maintains, that the *consistency* of the axiom system is *by no means assumed*. This consistency was in any case proved earlier and is no more thrown into doubt than algebra is thrown into doubt by attempts at "squaring the circle".

§31. Next Baer says: "N is indeed not postulated but constructed!" But how can one construct a thing which has these inconsistent properties? Such a thing cannot exist at all. To think that impossible things can be constructed is exactly what leads to the antinomies.

What does it mean to "construct"? A *craftsman* may construct many things but ultimately their number is *limited*. He cannot construct arbitrarily many while always being able to produce one more.

In *geometry* the word "constructible" has a definite meaning. This meaning presupposes that geometry, or at least the number system, is *already given*. It still must be checked whether a required construction can be *performed*. Even the task of constructing a triangle given three sides is not always possible; certain *conditions* have to be satisfied.

Why should a "construction" always be feasible in *set theory*, even when it is contradictory? *Circle-free* constructions are indeed possible; but one must be on guard against non-satisfiable circles!

It is shown that there exist non-satisfiable rules of construction by the example given in my "Reply" [1928]: Write on a blackboard a number that is one greater than the largest number written on the board. Of course, it is meant that the "new" number shall be one greater than the greatest number written *formally* on the board. There does not exist a number which satisfies this requirement when the requirement itself is on the blackboard. One cannot claim that the "new" number is constructed!

Some might think that this example shows that the system of numbers written on the blackboard is always capable of extension. This does not hold either: As soon as the blackboard is full there cannot be any more.

§32. It is true that one cannot be prohibited from thinking of a thing **N** which possesses a certain relation to exactly all sets A for which $A \, \beta \, A$ does not hold. But this thing **N** is *not a set*, and the "new" relation is *not* the β-relation but another relation, say γ, which cannot be identified with β.

If one starts from the ϵ-relation which is inverse to β, one can introduce η as a new relation which is inverse to γ. Baer uses such a relation with systems, so that $A \, \eta \, \Sigma$ means that the set A belongs to the system Σ. One cannot always identify η with ∈ because not every system corresponds to a set.

In spite of this, one is at liberty to consider the system as an "individual thing", which in general need not be a set. The object **N** considered above is *not a set* but at most a *system*.

That the originally given primitive relation β must be carefully distinguished from the other relations *derived* from it was given special prominence in an additional note to the "On the Foundations of Set Theory, Part I" [1926b, §18]. Thus it is not permissible to identify the b-relation with the ∈-relation which is derived from it, or conversely, although both relations satisfy axiom I equally well.

The *same* objects can thus stand in *different* relations to one another: The fact that one relation can be derived from another does not prevent them from being *distinct*.

The subset relation is *derived* from the membership relation and cannot be interchanged with it. It is hardly surprising that such interchanges lead to error.

In the axiom system for sets only *one* relation β is given explicitly; it *cannot* be replaced by a *derived* relation.

§33. Hilbert's observation that in the absence of the Archimedean axiom, the axiom of completeness for geometry would constitute a contradiction is quoted by Baer. This was of course well known to me at the time of writing "On the Foundations of Set Theory, Part I" [1926b]. However, as I have already indicated in the "Reply" [1928], it is not to be taken literally. It deals with an *apparent* contradiction that can easily occur in set theory as well.

The statement that a system of points, lines and planes that satisfy Hilbert's axioms I – IV can always be extended, holds for the relatively simple systems that can be obtained from the usual geometry by means of certain successive extensions. Such a system can always be extended as long as one does *not* consider *all possible* extensions. But the union of *all possible* extensions gives rise to a system which can no longer be extended and which, therefore, satisfies the axiom of completeness.

In a similar way it can be shown in naive set theory that the set of all subsets of a set is of greater cardinality than the set itself. Thus, in *naive set theory* there cannot exist a greatest cardinal number. In *complete set theory*, which encompasses *all* pure sets, this statement no longer holds; there the set of all sets has the greatest cardinal number. This is only an *apparent* contradiction, and in geometry it is exactly the same.

Thus, *contrary* to the remark of Baer, the axiom of completeness *can be applied*. It is *necessary* for a complete theory having *no restrictions* on set formation.

The precise meaning of Hilbert's observation is that in geometry the Archimedean axiom ensures that the axiom of completeness is *already* satisfiable *in a circle-free domain*. It is of course quite obvious that Hilbert meant only domains of this kind, even though he did not specifically define them. In *these* domains his observation holds.

A restriction similar to that provided by the Archimedean axiom in geometry is, however, *not known* for set theory. Moreover, it is just this *unrestricted* set theory that is to be investigated here.

Thus, there really is a *fundamental difference* between the application of the axiom of completeness in Euclidean geometry and its application in set theory. This axiom and also the essential use of the concept "all" are *necessary* for completeness in set theory. In geometry, however, replacing the axiom of completeness by other postulates which do not presuppose the concept "all" is *conceivable*, at least in principle, though it is probably not feasible in practice.

For example, if the continuum is defined to be the set of all subsets of the system of natural numbers, then the concept "all" is used, but only in such a way that the result still remains *circle-free*. These subsets can in principle also be considered as being given individually. This also holds for the class of all sets; this latter necessarily presupposes the concept "all". Otherwise it could still be extended. The definition "all even prime numbers" can be replaced by the definition "the number two"; thus in this case the concept "all" is not necessary. On the other hand, "all sets" cannot just be listed. If the sets essential in themselves are excluded, then without the concept "all" only the circle-free sets are obtained, not all sets. But this is no reason for rejecting the "totality of all sets", it just cannot be obtained in a circle-free fashion.

Compare this with the remarks in §60 below.

§34.   If only axioms I and II are retained and the axiom of completeness is omitted, then the *extent* of the domain of sets under consideration is extremely arbitrary. Axioms I and II are satisfied in every complete system. One can thus proceed from an arbitrary set or also from an arbitrary class of sets and consider the sets essential in them. In either case the axioms are satisfied.

This has *nothing whatever to do* with the special investigations of Skolem [1922] into "set theoretic relativity" to which Baer refers [1928a, 539]. The cardinal numbers in these domains always retain their absolute meaning as was pointed out in "On the Foundations of

Set Theory, Part I" [1926b, §12]. They are *not* made *relative* by means of specially prescribed rules of construction; these do not arise here. That, however, is certainly what is meant by the relativity to which Baer refers.

A set is *uncountable* if there does *not exist* an enumeration of its elements and not because there is no enumeration obtainable by *special methods*. The class of *all* subsets of the natural number system is *absolutely* uncountable; it cannot be found in a countable domain.

If, in spite of this, one admits only those subsets which satisfy certain special conditions, then it can be that one obtains only countably many.   To say that this class is metaphorically uncountable is *unnecessary* and *misleading* from the absolute standpoint.

What is meant by Baer's concluding remark that "this set theoretic relativity is at least made safe 'upwards'" [1928a, 539]? This really is *meaningless*.

§35. It still has to be asked whether it is *practical* to *restrict* the domain of sets, perhaps by using rules of construction.

For the *investigation* and *clarification* of the set theoretic *antinomies* one must use the *total system of all* pure sets: It is just here that these difficulties really become apparent. It is very important to know that this system *exists*.

One needs this system to define *circle-free* sets. These sets form a system in which, without *unnecessary restrictions,* the usual axioms of set theory hold.

One also requires these sets in order to *prove* the fact that *infinitely many objects* and also *uncountable cardinal numbers* really *do exist*. This is the *only* means by which the existence of the *infinite number line* and the *continuum* can be assured.

The class of *circle-free* sets suffices for all *applications* of set theory.

Additional restrictions could be *useful* for many purposes, just as one may find it useful to restrict a numerical computation to a certain number of significant figures.

Number theory proper ought to deal with the totality of *all* numbers: Similarly *set theory* proper concerns the totality of *all* sets.

It should be particularly stressed that sets are only a *generalization* of the natural numbers: These latter possess the relation β to their "predecessor", either to exactly *one* other natural number or to none. Sets have this relation β to their "elements", to arbitrarily many sets.

§36. Everything written in my "Reply" [1928] to Baer [1928a] remains completely valid. One should refer to it rather than have these points repeated here.

In his observations [1928b] to this reply Baer disregarded almost everything and only two points were picked out. With regard to both of these he is wrong once again.

The assumption that the construction of the set N is of a "circular nature" in certain circumstances (these Baer suppresses) certainly does not stem from a misunderstanding on my part.

Baer has not placed restrictions on the system of sets satisfying axioms I and II. Now, if this system Σ contains all sets, which is not prohibited, then, precisely through the fact that it is to be a set, the allegedly new set N would certainly have to belong to the system Σ. That is the circle.

The arguments of Baer are a typical example of how one can "prove" false propositions with a paradox which is not understood.

Should the "new" thing N (or better **N**) be a *class* rather than a *set*, then the circle disappears and with it the contradiction. But then there is no extension of the system of sets. If, however, one requires that N be a *set*, then this set *must*, in as much as Σ contains *all* sets, belong to the "already existing sets in the proposed set system Σ", even if "it is explicitly demonstrated" that this is not possible. Thus in this case there is a non-satisfiable circle and a contradiction; such a set N *does not exist*!

As with the usual paradox of the "set of all sets which do not contain themselves as elements", which we are actually dealing with here, it is "explicitly demonstrated" that it cannot contain itself. In spite of this it would have to contain itself and therefore it *cannot exist*.

§37. The second of Baer's remarks concerns Hilbert's argument, quoted in [1928a, 539], on the axiom of completeness (see above §33). In order to attack my explanation Baer maintains that "*every* real field can be extended to form a larger real field adjoining a transcendental element - although this field need not be non-Archimedean." [1928b]

This, however, *does not hold*! Just as a greatest transfinite cardinal number is obtained from the union of all transfinite ordinal numbers (see §4 above), so also a real field, the *greatest* real field, is obtained from the union of all transcendental extensions. No further transcendental elements can be adjoined to this: To suppose as much would result in an unsatisfiable circle, and therefore with a contradiction.

In this case, one does not advocate a *restriction* to *circle-free* objects.

Obviously, *previous* research *tacitly assumed* that all constructions are circle-free, since it was not realized that circular contructions could safely be carried out using the concept "all". One must be clear about what is meant and one must define what circle-free constructions are in *general*. Otherwise one just proves *falsehoods*.

§38. Fraenkel's reviews [1928a, 1928b, 1928c] which are quoted below refer to the discussion between Baer and myself in the following way.

> Baer [1928a]: "A criticism of the crucial point of Finsler's "On the Foundations of Set Theory, Part I" [1926b], namely it is to be shown that each consistent model of a set theory of Finsler's kind is always still capable of extension - in contradiction to the 'axiom of completeness' given there."

> Finsler [1928]: "Attempt at refutation of the aforesaid criticism".

> Baer [1928b]: "Countercriticism of the previously mentioned reply."

As has been demonstrated in detail in §15 - §35, the criticism of Baer [1928a] is *totally untenable*, and also *false* in its crucial contention.

The "Reply" [1928], as shown in §36 and §37, was not invalidated by the erroneous countercriticism of Baer [1928b], and is, as has been made abundantly clear in the course of the preceding, much more than a mere "attempt" at refuting of Baer [1928a].

It appears that Fraenkel has given insufficient attention to my reply [1928]. At least, he has not read "On the Foundations of Set Theory, Part I" [1926b] accurately; this even shows in his text ([1923, 200] and [1928d, 289]) where he retains the statement that in Zermelo [1908] the axiom of choice (in its usual formulation!) can be stated without "essential" mention of disjointness of the given sets. This indicates unfamiliarity with the counterexample, consisting of the set $\{\{a\}, \{b\}, \{a, b\}\}$, in §17 of "On the Foundations of Set Theory, Part I" [1926b].

§39. I can assure Fraenkel that I did *not take the idea* of investigating *pure sets*, whose elements are themselves again only pure sets, *from him*. At the time of my inaugural lecture at Cologne in the year 1923 where this idea was expressed and established (Finsler [1925]), his investigations into these matters were *not known to me*. A few years before I had intimated to Bernays, for instance, about the idea of adopting the system of pure sets as the foundation for set theory.

In passing it might also be mentioned that "pure sets" had been used already for the theory of ordinal numbers by Zermelo. In addition, Fraenkel has *not* introduced a *name* for them.

The *really important point* is not that one considers only pure sets, however, but it is the insight that this is the *sole* restriction that is needed in order to obtain a *consistent set theory*. One obtains a *clearly defined* system of sets which contains *all* pure sets and which thus corresponds to the *unique system* of *all* natural numbers (see Finsler [1925]). Even to this day Fraenkel is still infinitely far removed from this insight.

Chapter 3. A Review

§40. T. Skolem has reviewed "On the Foundations of Set Theory, Part I" [1926b] in [1926]. This review has been *reproduced* here *completely* to allow a full commentary.

It begins thus.

> P. Finsler, On the Foundations of Set Theory, Part I: Sets and Their Axioms [1926b].
> This paper contains an attempt at founding set theory in such a way that on the one hand it does not lead to the antinomies and on the other it constitutes an absolute and uniquely determined theory.

That is *correct*, only the attempt has not failed as Skolem later suggests but has indeed *succeeded*.

§41. It continues.

> The attitude of the author emerges sharply already in the introduction, in that he says there that for the truth or falsity of a mathematical proposition it is quite

irrelevant whether we can prove or refute it with our human means.

That is *not* exactly what I *said*! A mathematical proposition is self-evidently true if we can prove it by our human means and it is false if we can refute it. This is *certainly not irrelevant*!

A mathematical proposition does not, however, *become* true only when we can prove it and it does not *become* false only when we can disprove it; it is true or false even if we do not know which is the case.

Some might say that the proposition is true "for us" only when we have proved it. That, however, means something different, namely that we *realize* that the statement is true only when it has been proved.

We are investigating pure mathematics; it is independent of the course of time and of our limited resources.

§42.

> Many will find that what is essentially meant by an undecidable proposition being "true" or "false" is very unclear.

Skolem *cannot have taken* up the *reference* to "Formal Proofs and Decidability" [1926a] that was given at the relevant place in "On the Foundations" [1926b, Introduction]. In that paper it was shown that there are false propositions (and others that are true) that are *formally undecidable*, but which in spite of this can be *seen* to be actually *false* (just as with others, that they are true). It is clear from this what is meant by absolute truth or falsehood of formally undecidable propositions (it is always formally undecidable propositions that Skolem has in mind). The question as to whether there also exist *absolutely undecidable* propositions has been investigated later in [1944].

The paper "On the Foundations" [1926b] *presupposes* the earlier paper [1926a]. How can anyone assess the situation correctly if he lacks one of the basic underlying assumptions?

We do not know to this day whether Euler's constant is rational or irrational, and it is conceivable that we shall never be able to decide this matter. In spite of this, we say it is rational if it is equal to the quotient of two rational numbers and otherwise irrational: The law of the excluded middle holds. In "On the Foundations" this law is *presupposed* to be part of basic mathematics; this means that only those things that satisfy it are investigated, that is, those *whose existence is unambiguous*. Whether we can *personally* decide does not matter.

It was also clearly stated in "On the Foundations" [1926b, Introduction] that those investigations for which the law of the excluded middle is rejected are not being considered. If Skolem wants to judge the work, then he is obliged to adopt this standpoint, that is, he must make *the same assumptions* and not speak about other things which have absolutely nothing to do with the work in question.

In pure mathematics it is usually supposed that the meaning of "true" and "false" is known.

§43.

> In chapter I, §1 he discusses first of all the false assumption which leads to Russell's antinomy, i.e., the assumption that one can so reason with a domain of things that any of the things whatever can be collected together into a set, which is again a thing of the domain.

This does not, however, mean that one is not allowed to speak of the system of all sets. The only thing that is forbidden is the assumption that every class N of sets corresponds to a set *N*. There really are classes that do not correspond to sets.

> In §2 he speaks of circular definitions. The author's later treatment of sets of certain objects as not being identical to their totality, or aggregate, but only things which are associated with the objects, appears to the reviewer to be only a playing with words; for, every collection of objects can be viewed as such a correlated thing, quite irrespective of whether it is called a set, totality or aggregate.

As usual, the *objects themselves* are collected into a "totality" or "aggregate". In a thorough set theory a *set* of objects must be a *single* thing; this is thus something *quite different*. I have certainly made this difference *very clear* in "On the Foundations". It is incomprehensible to me that anyone can call this "playing with words".

If *one* set possesses *two* elements, then it is certainly not the case that one thing is identical to two things, with the consequence that one equals two. If matters like this are confused, then of course one will arrive at contradictions and antinomies, rather than a useful set theory.

It is of no significance that *linguistically* the singular is used when *the* class of elements of a set is mentioned. It is all a question of what is *meant* by this. In this case it is *not* the *collection* as such which is *meant* but the *elements* themselves and indeed all of them without exception. If the class (or totality) of the pupils in a school has passed an exam, then it is not an abstract concept which has passed the exam but each individual has passed.

The heading of §3 of "On the Foundations" was "Sets and Classes". In §4 *system of sets* were introduced where the *sets* are *objects*; the *systems*, however, are *collections* of sets.

The distinction between "set" and "class" is quite common today in set theory. According to Weyl [1946, 11] "the introduction of classes [...] is due to Fraenkel, von Neumann, Bernays and others."

§44. Here I want to interpolate the following remark: According to Cantor a set is, put briefly, the collecting of well determined objects into a whole. In [1925] I commented on the expression: "thus the collecting itself".

In [1928d, 13] Fraenkel says with regard to Cantor's definition: "Naturally it is not the *act* of collecting but the *outcome* of this act that is meant." Is that really so natural? In any case Cantor did not say this. In a *very naive* set theory one will, to be sure, at first think of the *result* of the collecting. But what does one get as a result if one collects together, for example, the number 1? Well, the number 1 itself, of course! In spite of this one says in set theory that one must not confuse a set which contains only *one* element with this element. And what does one get if one collects together nothing at all? Well surely nothing at all! How then does one come to the empty set?

The observation of Fraenkel can only serve to *complicate* insight into the foundations of set theory and above all into the essence of the paradoxes. Why should one not understand by the concept "set" the *operation* of collecting elements together? Then everything is clear. The result of an operation cannot be obtained without carrying out the operation; therefore, this latter is certainly necessary. If, however, because of a non-satisfiable circle, the operation cannot be performed, then there will not be any result either. If, in spite of this, the result is *postulated*, then one has an *antinomy*.

§45.

In chapter II the author sets out his axioms. According to axiom I it shall always "be decided" whether $M \beta N$ (which means N is an element of M) holds or not.

Of course, "be decided" is an application of the law of the excluded middle; it is not required that we can always come to the decision ourselves.

It has been explained in §16 above, why β rather than its converse was taken as the *basic relation*.

> According to axiom II, M and N are to be identical whenever all sets essential in M (this means elements of M, elements of elements of M, etc.) can be associated with the sets essential in N in a one-to-one and β-relation preserving fashion.

That the elements of M must be associated with the elements of N, can very well be considered as self-evident here.

§46.

> Axiom III is an axiom of completeness; it states that the domain of sets under consideration shall be maximal. It appears, however, to be clear that every domain which satisfies axioms I and II must be capable of extension, unless one wants to prohibit the formation of new things, for example, if one wants to prohibit the aggregates formed out of things of the domain from being regarded as new things, which appears to be senseless (see Baer [1928a] and Fraenkel [1928a]).

It has already been indicated here the value that can be attached to the remarks of Baer [1928a] and the review of Fraenkel [1928a]. It appears, however, as though the axiom of completeness presents particular difficulties. Let us turn to it once again.

For every natural number there exists a greater natural number so that in counting one can always say "I am not finished". More exactly, whatever the last number one has not reached the end. In spite of this it is *not* permissible to *forbid* one from speaking of *all* natural numbers, that is, of all that there are, or of all that there can consistently be. Such a prohibition would really be *unfounded* and therefore "senseless". We do not *have to* count up to a number for it to exist; that would be a *very unfair requirement*.

It is precisely the same with sets. From any ordinary set whatever, for instance from the empty set, one can always proceed to form further sets so that also here one apparently "never comes to an end". In spite of this *no* one can be *prohibited* from speaking of *all* pure sets, that is of all that there are, of all that exist consistently.

Such a prohibition would be equally *unfounded* and therefore "senseless".

Pure sets are indeed only a *generalization* of the natural numbers (see §35). Whereas the latter always have only *one* predecessor, a set possesses *arbitrarily many* "predecessors", namely its elements. That is in principle the only difference.

A "set" is always to be understood here as a "pure set"; the elements of a pure set are only sets. These sets are, *just as are the natural numbers*, only *ideal objects* which stand in a certain relation to one another.

§47. The domain of all sets can no more be extended than the natural numbers can; consequently it *does satisfy the axiom of completeness*.

In addition to *all* natural numbers there cannot still be "formed" or "constructed" still one "new natural number"; the concept would be contradictory and the formation of *impossible, contradictory* "things", is *self-evidently prohibited* because it simply is not feasible!

In exactly the same way one cannot "form" or "construct" over and above all sets a "new set"; this would also be a contradictory concept. The formation of *impossible*, that is *contradictory* "things", is *self-evidently prohibited* also.

If to each natural number a successor can be "formed", why cannot one also form a successor to all natural numbers? Because it constitutes a contradiction! One can certainly think of a "new thing" and decide that it might follow all natural numbers; one could also, if one wants to, describe it as an ordinal number and symbolize it by "ω". However, it is *not a natural number*. A natural number which is the successor to all natural numbers *does not exist*!

The formation of new things is thus not arbitrarily forbidden, but it is only possible when there are still new things, that is, when no contradiction arises in the "formation". It is *not permitted*, however, to regard a *new* thing as one of the *old* things; that is a contradiction!

So it is *not forbidden* to view the aggregates formed from things of the domain as new things, as long as one does not say that these "new things" are old things, that is, to say that they are sets.

If the system of all sets which do not contain themselves "is regarded as a new thing", then this "new thing" is a *class* or a *system* and *not a set*; otherwise it would certainly not be a "new" thing. This class is just as little a set as is the ordinal number ω a natural number. The domain of *all sets cannot be extended* by adding new *sets*; it already is *maximal*.

§48. The domain of *all rational numbers* can be extended to the domain of *all real numbers* by describing certain classes of rational numbers as real numbers. To every rational number there corresponds a unique real number; conversely, however, there exist real numbers to which no rational number corresponds. These cannot be called rational numbers; otherwise contradictions will occur.

In exactly the same way, the domain of *all sets cannot be* extended to include the domain of *all classes of sets*. To each set there corresponds a unique class of sets, namely the class of its elements. Conversely, however, there are classes of sets which do not correspond to a single set; contradictions arise if they are designated as sets. An example of this is the class of all sets that do not contain themselves.

The irrational numbers distinguish themselves from the rational numbers by the fact that they cannot be represented as the quotient of two integral rational numbers. Similarly, the classes to which no set corresponds distinguish themselves from sets by the fact that they cannot be represented by the *unique* relation β. Sets are determined solely by means of the relation β. In order to obtain a system or class of sets it is first necessary to have the sets, that is, to have the relation β. One could then produce a new relation γ which specifies the membership of sets in a class. The relation γ has no meaning without the relation β. Only in special cases could it be replaced by the relation β, namely, when the class forms a set. Similarly a real number, too, as a class of rational numbers, can be replaced by a rational number only in special cases; with irrational numbers this does not work.

The axiom of completeness is not the least invalidated by the fact that the domain of classes extends beyond the sets. The axiom of completeness refers only to *sets*, that is to those things given by the primitive relation b (see §32). *A larger domain of sets does not exist!*

§49. [Skolem continues.]

Further, it is clear that the requirement that the system under consideration be the largest which satisfies I and II can only have an absolute meaning if the totality of all systems is already uniquely determined by other means; but then the problem as to whether there occurs a largest system within this totality would have to be solved first of all.

The totality of all systems which satisfy axioms I and II is actually determined by means of these axioms; it constitutes all that there are, that is, all that are consistently possible.

The problem as to whether there is a largest system is solved by showing that these systems can be united into a largest, by identifying isomorphic sets of different systems. This was carried out in §9 and §11 of "On the Foundations of Set Theory, Part I" [1926b].

§50.

In a more detailed discussion of axiom III in §10 the author says that a set exists whenever the assumption of its existence does not lead to a contradiction with I and II. It appears, however, to the present reviewer to be quite illogical to establish the existence of things by such a definition within a theory where the things are not isolated logical structures but stand in the most manifold relations to one another.

That a set M, defined in any way whatever, "exists" means in this context, exactly as was stated in §10, that there exists in the system Σ of all sets one set which satisfies the definition. The supposition that the existence of M does not contradict the first two axioms means, as was also stated here, that there is a system satisfying the first two axioms containing a set M that satisfies the definition.

If no such system were to exist, then the assumption that there is such a set would already contradict the first two axioms. If, however, one such system does exist, then the set M must also belong to the total system Σ, since this, as the union of all systems which satisfy I and II, must also contain the particular system, and with it also the set M. This is definitely not "illogical".

Sets stand "in the most manifold relations to one another" in just the same way as do, for instance, the real numbers. In defining a fixed real number, one does not need to know these relations; it suffices that the definition in itself be correct. That is, it unambiguously and uniquely determines a real number. It is exactly the same with sets.

Sets can be considered as "isolated, logical structures" just as well as the real numbers, or, what is more to the point, just as well as the natural numbers.

A real number can be defined by means of a fixed system of rational numbers. Similarly a set can be defined by means of the system of its elements together with the sets essential in them. Not every system of rational numbers defines a real number: Not every system of sets defines a set. Certain conditions must still be satisfied; in the latter case there is only one condition: that the existence of such a set does not contradict axioms I and II.

§51.

> If the assumption of the existence of M and the non-existence of N gives no contradiction and also that of the existence of N with the non-existence of M, while the assumption of the existence of M as well as N gives a contradiction, what should really exist? How would it stand with the non-ambiguity of the theory?

How could something like this be possible? How could the existence of one set be dependent upon the non-existence of another set? The *existence* of a set is an *absolute*, and *not a conditional*, property. If a set is defined in a definite way, for instance by specifying its elements, then the assumption of its existence either stands in contradiction to axioms I and II or it does not. Nothing else is possible!

If a set exists, then of course the sets essential in it must exist; it is of no concern what other sets there may be.

It is not maintained that in every case we must *know* or be able to *decide* whether a well-defined set exists or not, nor do we know whether every well-defined real number is rational or irrational.

It can happen that a system which satisfies axioms I and II contains the set M and not the set N, another may contain N and not M. When we identify isomorphic sets, the two systems can be united to yield a system which still satisfies the axioms and contains both M and N. No contradictions can arise from the identification of isomorphic sets; there are no grounds for not identifying them.

Therefore, if a set M and a set N separately exist in an absolute sense, then *both of them* exist!

For these reasons the uniqueness and absence of ambiguity of the theory is also assured.

§52.

> And what moreover does "contradiction" mean in Finsler's theory? According to the explanation in the introduction a "contradiction" does not need to be a "demonstrable contradiction". A more exact definition is then, however, lacking.

Why should "contradiction" mean something special in my theory? The word "contradiction" means exactly what it implies and what it has always meant in classical mathematics. A contradiction need be demonstrable just as little as a murder needs to be provable.

A special definition would be superfluous here; a merely *formal* definition would be completely out of context. We refer to what has been said in §12.

A real number is irrational if and only if the assumption that it is rational contains a contradiction. In exactly the same way a class of sets does *not* form a set if and only if the assumption that there is a corresponding set contains a contradiction.

In both cases it is not necessary that *we* can *demonstrate* the contradiction. The law of the excluded middle holds; the assumption is either contradictory or it is not.

If the supposition that a certain positive real number is rational contains *no* contradiction, then *there exist* two natural numbers whose quotient is equal to this number, whether we can find them or not. If there were no such numbers then the assumption would be contradictory, because it would contradict the facts.

If the supposition that a certain class of sets forms a set contains *no* contradiction then *there exists* a set which contains exactly the sets of this class. Only sets belong to the total system Σ; this is, therefore, uniquely and consistently determined.

The consistency of geometry rests upon that of arithmetic, arithmetic rests on set theory, set theory is based on logic. Logic is already consistent; otherwise it would not be logic.

It is clear that this is always *absolute consistency* and not merely formal consistency; similarly the logic is *absolute logic* and not only formal logic.

§53. [Returning to Skolem]

In §9 the author asserts that quite unrestricted unions and intersections of systems of sets can be formed. Here, he has no scruples about the problem relating to non-predicative definitions so that he does not find it necessary to make distinctions between types.

Already in §4 of "On the Foundations of Set Theory, Part I" [1926b] it was observed: "From this axiomatic point of view sets are things in themselves rather than collections; it will soon become apparent that we can collect sets together without risking the danger of circular constructions." This means that unrestricted unions and intersections of systems can be formed. A set belongs to the union of given systems if it belongs to at least one of them; it belongs to their intersection if it belongs to all of them. That is quite unambiguous.

The distinction between sets and systems of sets can, if one wants, be viewed as a distinction between levels. The distinction between sets and classes is all that is required here.

Systems must always be well-defined and consistent; whether this can be achieved with non-predicative definitions need not concern us here. One must examine special definitions case by case. A system of sets is given whenever it is unambiguously determined whether each set belongs to the system or not.

An example of a non-predicative definition given in "On the Foundations of Set Theory, Part I" [1926b, §15] will be discussed in greater detail under §62.

§54. [Skolem next turns to the concept of circularity.]

> In chapter 3, which deals with the formation of sets, the concepts "circle-free" and "circular" are introduced. A set M is said to be circle-free whenever it together with all sets essential in it are independent of the concept "circle-free", which means that the definition of M always yields the same sets irrespective of which sets are characterized as circle-free.

Prior to this we had already excluded those sets in which a set which is essential in itself is essential, that is, all sets which are already "crudely" circular, including those sets that contain themselves.

Strictly speaking Skolem's review ought to read: " [...] are independent, which means *can* be so defined that the definition always yields the same set irrespective of which sets are characterized as circle-free". A circle-free set can also be defined in a circular way. In addition, a set in which a circular set is essential is circular, even if the set as a whole is independent of the concept "circle-free".

> Here it is extraordinary that the concept "circle-free" should have no connection with the original (primitive or fundamental) relation β.

The concept "circle-free" relates to sets, and these are determined by the primitive relation β. In as much as this is so there is certainly a connection with this relation.

§55. The concept "circle-free" is of the greatest importance as will be specially shown in §61. It may be misunderstood by those who do not see that one can speak without contradiction of all sets. A few comments and examples will nevertheless be given.

Call a set "cyclic" if some set essential in it is essential in itself. *Exclude* all cyclic sets from consideration. One of the remaining sets is circle-free if it *and everything essential in it* is definable without using the concept circle-free. The empty set is circle-free, since it can be defined as a "set without elements". Similarly the unit set, which contains only the empty set, is circle-free. Continuing in this way one can form many more circle-free sets. These can be called "unproblematic" sets.

It is not possible, however, to simply consider all non-cyclic sets as circle-free. This would not serve to eliminate the paradoxes which rest on concealed circles.

A "set of all non-cyclic sets" cannot exist: If it were not cyclic, then it would have to contain itself and therefore it would be cyclic. Should it be cyclic on the other hand, then a set essential in itself would have to be essential in it, which is likewise impossible.

Genuinely circle-free sets, in whose definition no hidden circle is contained, certainly should be capable of being collected together into a "new" set. What would there be to prevent this? A circle would cause an obstruction. Were the circle to appear in the membership relation, then the "new" set would have to contain itself and would be an "old" set.

Such a circle must necessarily be contained in the definition of the set being sought, as its elements are all circle-free sets. But then the new set, if it exists, could not be considered circle-free; it would have to be circular. As a *circular* set it is, however, really a *new* set; it could not belong to the old, circle-free sets. Thus, if a well defined set of circle-free sets has only circular definitions then it is circular. But nothing need stand in the way of its existence as a circular set. In that event the circle is satisfiable. If, however, a set of circle-free sets can be defined in a circle-free fashion, then it is circle-free.

In particular, the "set of all circle-free sets" is a circular set; otherwise it would have to contain itself. This is a simple example of a circular set which is not "crudely" circular. The concept "circle-free" really does occur in its definition.

One might think that this set, as a "set of all sets whose definitions are free of concealed circles" could also be defined explicitly, that is, in a circle-free way. In actual fact, however, this definition also contains a concealed circle, in that it refers to concealed circles. Similarly, as everyone knows, it is not possible to think about the absence of white elephants without thinking about white elephants.

§56. A set of circle-free sets is therefore circular whenever its definition necessarily contains the concept "circle-free", which can only be defined in a circular way; circumlocutions involving this concept would have to be replaced by the concept itself. If the set is circle-free, then it must be definable without reference to this concept, that is, independently of it. It is apparent that this condition suffices to ensure that a set can be considered circle-free.

It should not be construed from this that *we* can always define such a set without the use of a circle. Even with the set of all natural numbers, or any infinite set, this is not the case, since *we* cannot enumerate all the elements separately. A circle-free set can, in spite of this, be *thought* of as being given by a mere presentation of its elements, without thereby giving rise to a logical contradiction. With circular sets whose elements are circle-free this is not possible, because here the concept "circle-free" must necessarily occur in the definition.

That there exist systems of sets which cannot be given by the mere presentation of their sets, is also shown by the system of all sets. Here the definition necessarily must contain the concept "all", otherwise there would be no reason why the system could not be extended. Compare this with §60.

§57. It remains to be asked how one can recognize that the concept "circle-free" *necessarily* must occur in the definition of a set. A set could be circular even if all of its elements are circle-free. The concept "circle-free" might only *appear* to be necessary. For example, the empty set, which possesses neither a circle-free nor a circular element, is circle-free.

In order to decide whether a function $f(x)$ actually, or only apparently, contains the variable x, which could be fixed by the relation $x = f(x)$, one *varies* x in order to see whether $f(x)$ always has the same value or not. In the latter case the independence of the function $f(x)$ from the value x is assured.

In a similar way one can decide whether the definition of a set really or only apparently contains the concept "circle-free": One allows this concept to *vary* and then checks whether or not on the basis of the definition it is the same set or not. If the set changes, the dependence is assured.

Varying the concept is accomplished as follows: One *designates* arbitrary sets as circle-free and the others as circular, even if this does not coincide with the final determination.

If *no* set is designated as circle-free, then the "set of all circle-free sets" is the empty set. If only the empty set is designated as circle-free, then the set of all circle-free sets is the unit set, which is something different. This shows that the definition given is actually dependent upon the concept "circle-free".

This test can thus be carried out even when the final determination of the concept is still completely unknown.

If the definition of a set is dependent upon the concept "circle-free", then the set is still not necessarily circular; there could exist another definition for the same set which does not contain this concept. Thus the "set of all circle-free empty sets" is the unit set and is therefore circle-free, although the definition, as given, is not independent of the concept "circle-free". Were all sets to be designated as circular, then this definition gives the empty set; that is something different.

Thus, as given in "On the Foundations of Set Theory, Part I" [1926b, §13] a set shall be said to be "independent of the concept 'circle-free'" whenever it *can be so defined* that the definition always yields the same set, irrespective of which sets are designated as circle-free.

A given non-circular set *is* then *circle-free* if not only it, but also every set essential in it, is independent of the concept "circle-free".

That this definition is all right in spite of its circularity was shown in §14 of "On the Foundations". The present discussion serves only as an elucidation.

§**58**. Skolem now continues.

> What is more, the double application of the concept "circle-free", once arbitrarily varied in order to investigate the effect on the definition, and once constant or definitive (final), is of course well suited to leading into confusion. In order to make matters clear one would have to state explicitly on the one hand "varying circle-free" and on the other "definitively circle-free". If, however, one does this, then it appears that one comes to other conclusions than those of the author.

There is a difference between designating a set *definitively* as *circle-free*, or whether one designates it "definitively circle-free". If such things are confounded then of course confusion will arise.

The sets *designated definitively* circle-free *are* circle-free sets whereas the *provisional designation* can be varied.

In order to derive a value x from a relation of the form $x = f(x)$, one can at first *vary* x until one finds a value which satisfies the relation; this latter is then the *definitive* value of x.

If, however, in order to avoid the circle, or "in order to make matters clear" the "x on the right hand side" and the "x occurring on the left hand side" are denoted by y and z respectively, then a relation $y = f(z)$ is obtained, from which, in general, neither y nor z can be determined. Similarly it is not permissible to replace "circle-free" in one instance by "varying circle-free" and in another by "definitively circle-free"; otherwise one might obtain some other result or none at all.

§59.

Proposition 11 states that the class (totality) of all circle-free sets is a circular set.

More exactly, proposition 11 [1926b, §15] states that the set of all circle-free sets exists and is circular; this means that the corresponding class *forms* a circular set.

It is clear, however, that the class (totality) of the ultimately circle-free sets must again be an ultimately circle-free set, in so far as it is a set at all; for the arbitrarily varied, provisional distribution of the labels "circle-free" and "circular" can indeed have no bearing on the existence of the ultimately circle-free set.

It is self-evident that the provisional distribution of labels can have *no bearing* upon the *existence* of any set whatever; it can only serve to *test* whether a well defined set is circle-free or circular.

It is difficult to say what Skolem *meant* with his mode of expression.

Perhaps he meant the following: The ultimately circle-free sets are independent of the concept "circle-free"; the arbitrarily varied provisional distribution of the labels therefore plays no role whatever in their definition; they are unambiguously determined without this. Now this does, indeed, hold for *every, separate* circle-free set. Their *totality* is *uniquely determined*; but the unique determination does not imply that it is circle-free. The class of circle-free sets as a whole cannot be defined without reference to the concept "circle-free"; the definition is dependent upon this concept and is therefore *circular*. If, *for a test*, the labels "circle-free" and "circular" are distributed differently, then the definition which depends upon these labels describing the "totality of all circle-free sets" will yield *different* classes, even if the *ultimate* class is uniquely determined.

Similarly, the definition of the *set* of all circle-free sets is dependent upon the concept "circle-free", and this set is therefore *circular*; from the *test* one obtains *different* sets, but *ultimately* only one *unique* set.

It does not help at all to speak of the "class of the ultimately circle-free sets" instead of the class of circle-free sets. The "ultimately circle-free sets" are the circle-free sets and nothing else; one has exactly the same collection.

If, however, one wants to *retain* the expression "ultimately circle-free" and *with it* test whether ultimately circle-free sets together form a circle-free set, then one has to vary this concept. One therefore *designates* some sets as "ultimately circle-free", which in reality are not, and conversely, just as one does when *designating* sets as circle-free. The result is again the same. What comes into question here is not how the concept is named but how it is defined. This does not of course exclude the fact that some *nomenclature* is *bad*.

If, however, the expression "ultimately circle-free" means that this concept *must not be varied* and also is not to be equated with the "circle-free" already defined by means of the variation, then it must be asked, whether it is to be defined at all. Other definitions can of course lead to other results. If Skolem defines the concept in such a way that the aggregate of "ultimately circle-free sets", provided it is a set, would be circle-free, then this would have to contain itself and, being a non-circular set, could not exist. This, however, shows only that such a definition is *utterly useless*.

§60. It may be conspicuous that there are classes of sets which can only be defined with reference to a fixed logical concept. Now that is a fact with which one must comply.

This fact is indeed unusual, but it does not occur only in connection with the concept "circle-free"; it is also to be found in the concept "all" (see §56).

If the concept of a set is unambiguously defined, then so is that of the class of all sets; to reject it as such has no justification.

The class of all sets cannot be defined, however, without reference to the concept "all"; that is, not in such a way that the sets can be thought of as being produced singly, without having used this concept. Otherwise new sets could still be found without giving rise to a contradiction; and therefore one would never have all sets. If, however, one speaks of the class of all sets and so uses the concept "all", then there cannot exist any more new sets, for the existence of one would constitute a direct contradiction.

Again, it is also the concept and not its name which matters. The totality of the sets which do not contain themselves is the totality of *all* such sets; the concept "all" is inherent already in the word "totality". In the axiomatic representation, the concept "all" lies in the axiom of completeness.

§61. The concept "*all*" is a well known logical concept which is of the greatest significance for mathematics. If one were not allowed to speak of *all* natural numbers, of *all* real numbers, of *all* sets, then one could never have clarity in mathematics.

The concept "*infinite*" is also a well known concept, which is of the greatest importance, and which receives its essential, strictly logical meaning in mathematics. Restriction to the finite, in particular to finite sets, would exclude the greater part of mathematics.

The concept "*circle-free*" is similarly a logical concept, which relates in particular to sets and so to the whole of mathematics; it is of the greatest significance for the *foundation* of mathematics, especially the *infinite* in mathematics. Without this concept, one would not succeed in securing the existence of the infinite and of the higher cardinal numbers.

Now, this is of course a *new* logical concept. Is it not allowed that something really *new* can turn up in logic, except for *formal* methods, which have surely been elaborated in the richest measure? Logic itself is invariant; but our *knowledge* of logic can change. It is certainly clear that one *must* have something new when one sees that the methods hitherto available no longer suffice to reach the infinite. What one truly needs is not something arbitrarily or artificially constructed but something which emerges from the nature of the subject and which suffices for it.

A real access to the infinite: does that mean nothing? Why should one struggle against it?

The correctness and utility of a logical concept are independent of whether or not some people fail to understand it, or do not want to understand it. A *tenable objection* against the concept "circle-free" has, to my knowledge, *never* been raised.

As soon as one realizes what the essential *difference* is between the usual sets, as they are continually used in analysis and geometry, and the paradoxical sets or classes of sets (especially the classes to which no set corresponds) then one is necessarily lead to the concept "circle-free". Why can the set of all natural numbers be formed, but not the set of all ordinal numbers? The reason is that the

The *finite* ordinal numbers are circle-free; the set of the *finite* ordinal numbers can be formed and is likewise circle-free, but *not finite*. The set of all *countable* ordinal numbers can be formed and is circle-free but *not countable*. There are still many more circle-free ordinal numbers. The set of all *circle-free* ordinal numbers also exists, but it is *not circle-free*. Beyond this there are still further circular ordinal numbers. If there were a set of all ordinal numbers that *exist*, then it would have to be *non-existent* and thus could not be formed. It is not surprising that the set of all ordinal numbers, circle-free *and circular* together, cannot be formed because of the circular ones.

If someone believes that this difference between "usual" and "paradoxical" sets can be defined or determined in some other way without excluding useful sets, let him try. *It is, however, not enough* simply to leave this difference as it is without a definition. Well known assertions, such as the non-existence of a greatest cardinal number, hold for the circle-free sets, but not for arbitrary sets and must therefore be formulated accordingly. It is, however, not sufficient to *declare* a few examples of sets non-paradoxical and therefore admissible, especially if one does not at all know whether they really *are*.

As long as one cannot show that there are infinitely many things, *every infinite set* must appear as *paradoxical*. If, however, these are excluded entirely, then there is very, very little left.

In "On the Foundations of Set Theory, Part I" [1926b, §18] it was shown that the set of all circle-free sets is an *infinite* set, that there thus exist infinite sets, and therefore also *infinitely many things*.

§62.

> In the definition of the "well-defined" class in §15 the expression "inherent contradiction" plays an essential role. Does this really mean anything?

The definition referred to reads as follows: A well-defined class is understood to be one which is defined *completely, unambiguously, and without inner contradictions*."

Two examples were given in §15 of "On the Foundations" to prepare for this definition and will be repeated here: "For example, one may designate the class of all sets which do not contain themselves as well-defined even though it does not form a set. In contrast, the class of all those elements of the unit set which are identical to the set which contain the elements of this class is obviously not a well-defined class, even though only the one circle-free element of the unit set appears in the definition."

The *first example* shows that "paradoxical" classes are to be seen as being well-defined classes, only so far as they are unambiguously and consistently defined.

From axiom I it follows that for each set M it is unambiguously decided whether it contains itself or not, that is, whether $M \beta M$ holds or not. In the system of all sets, therefore, the class of all sets which do not contain themselves is uniquely and consistently determined; it consists exactly of those sets M for which $M \beta M$ does not hold. This yields a *well-defined* class.

One could only obtain a *contradiction* by requiring that this class forms a *set*; that was not required in Part I. On the contrary, this shows that *not every well-defined class* of sets forms a *set*.

The *second example* was apparently too difficult for the reviewer; otherwise he would have remarked that the definition contains an *inner contradiction*.

Purely externally, according to the mere form of the definition, the contradiction is not evident. It is stated which sets belong to the class and which do not; by these means this class, which could be empty, appears to be determined.

The definition of this class contains, however, a reference to the class itself, which leads in this case to a non-satisfiable circle and ultimately to an "inner contradiction". There is no class which satisfies the given definition.

The contradiction emerges in the following way: The unit set contains the empty set as its sole element. A class of elements of the unit set can therefore either be empty or consist of the empty set. The set whose elements are the numbers of this class is thus either the empty set or the unit set. If the class being sought were empty, then it would, according to the definition, have to contain the empty set, since this is an element of the unit set. If it consisted of the empty set, however, then it would have to be empty because the unit set is not an element of the unit set. Both cases are thus impossible; the definition really does contain a contradiction.

The class being sought in this case is *not* a *well-defined* class. If it were empty, it would have to consist of the empty set; and if it consists of the empty set, then it would have to be empty. The empty set is not an "empty class".

This example also shows clearly that such contradictions are in no way bound up with the concept of the infinite, or even that of the non-denumerable, but can occur already with very simple finite sets and classes, as soon as one has to do with circular, that is "non-predicative", definitions. In order to exclude paradoxes it does *not* suffice to restrict considerations to *finite* sets; on the contrary it is necessary to exclude *contradictory definitions*.

The expression "inner contradiction" is moreover to be found already in the paper [1926a, §2], which immediately precedes "On the Foundations of Set Theory, Part I" but Skolem, as has already been remarked, has obviously not paid attention to it. It was shown there, that there are formally consistent axiom systems, lacking formally decidable contradictions, but which have recognizable contradictions anyway. An "inner contradiction" may adhere to such systems. On this point too, there was a clear reference in "On the Foundations" [1926b, Introduction].

It is self-evident that in the definition of a fixed class no external, directly visible, contradiction can be found; the class would otherwise not be completely and uniquely defined. With regard to implicit or "non-predicative" definitions the possible occurrences of inner contradictions must, however, be specially investigated, because here the *impression* can be given that the class being sought is well-defined.

Otherwise, however, the distinction between inner and outward contradictions is not of particular importance. This also follows from the additional remark that "in the place of Zermelo's concept 'definite' we shall have to put the concept *well-defined* or *consistent*." [1926b, §15] Further it is said: "A well-defined class of sets is the same as saying that the class forms a system (§7)." In §7 of [1926b] it is said of the subsystems of Σ: "Any such subsystem is then well-defined whenever, for each set, it is uniquely decided whether it belongs to the subsystem or not."

What has to be understood by a "well-defined class" is presented in a sufficiently clear way.

§63. In [1908, 263] Zermelo calls a proposition *definite* whenever "the basic relations of the domain decide, by virtue of the axioms and generally valid laws of logic, on the validity or invalidity of the proposition without arbitrariness."

Incidentally this definition was rejected as "too fuzzy". A rejection is justified in so far as the definition has no meaning, when one does not know whether a contradiction may follow from the axioms. This is, however, not shown in Zermelo's case.

Mostly the difficulty is sought in the expression "generally valid laws of logic". As long as one burdens logic with antinomies this is comprehensible. Instead of clearing up the antinomies and recognizing an absolute logic which is necessarily consistent, special "logical laws" have been sought whose consistency is *assumed*; an *attempt* has even been made to prove consistency in particular cases, but *without absolute logic* such proofs are certainly very precarious.

The concept "definite" then was replaced by a narrow concept that allows only countably many subsets to be formed from an infinite set. This collection then is *metaphorically characterized* as "non-denumerable".

Thus, the whole "formal set theory" can be "realized" in a countable domain, which therefore contains an insignificant quantity of the real sets which have higher powers. Even for this minute fragment it is still not shown that it is consistent.

Why does one not admit *all* subsets of a set, as long as they are consistently defined? Already the subsets of the natural numbers yield an *actually* uncountable system which contains *infinitely* much more than the whole of the contrived pseudo-theory.

If, however, one takes on such restrictions, how does it stand with the *satisfiability of Hilbert's axiom of completeness for geometry*? *Is Hilbert, because of this, now suddenly incorrect*? How could one then speak of plane or three dimensional Euclidean geometry without adding each time which of the arbitrarily restricted ones is meant; otherwise there is ambiguity.

§64. [Skolem continues:]

In §17 the axiom of choice is "proved" by invoking adoptability without contradiction. Thus everything goes through very easily.

In fact, a safe is much more easily opened by using the right key than by prying with a crow bar.

I really have not set out to be easy on myself, otherwise success would not have been possible. For all that, I have taken the trouble to *first* resolve the antinomies *correctly* and *then* build up the *whole* of set theory, that is, the theory of *pure* sets.

If one knows exactly where opposition is waiting, one can dispose of it there and need not fear further trouble in other places.

In §17 of [1926b] it is shown that the formation of a set using the axiom of choice *within the domain of circle-free sets* does not entail a circle and *therefore*, under the given assumption, is always possible. Zermelo's *axiom* of choice which postulates the existence of a choice *set* is thus *justified* here. In the domain of *all* sets it *fails*, however, as can be seen from an example (see Finsler [1941b]), since the elements "chosen" need not always form a set.

§65.

Finsler's work is of course a well meant attempt to save classical mathematics in its complete extent. One has got to say, however, that this attempt is misguided.

The final remark is mistaken.

II. Combinatorial Part

Introduction

The Finsler theory was intended as a way of defending classical mathematics from the urge to formalize it in reaction to the paradoxes of logic and set theory. The combinatorial fertility of this set theory seems at first to be one of its incidental products; but Finsler also believed that sets are generalized numbers and that the unsolved questions of set theory might be approached from that direction. So perhaps it is natural that his ideas should be stimulating from a combinatorial point of view.

There are two short papers of Finsler included here. They concern totally finite sets. Since totally finite sets are well-founded, we have also included an introduction to the rich combinatorical problems to be found among the non-well-founded sets. As a result this introduction to set combinatorics is of greater length than the two papers of Finsler that have been included. First of all, let us see what is actually in the two papers translated here.

The two papers in this section concern totally finite sets. Everything here is exactly the same whether it is understood to take place in Finsler set theory, ZF, or any set theory, in fact.

The totally finite sets are exactly the sets that are usually called "hereditarily finite sets". The reason that two different names are required is that the usual definition of hereditarily finite sets may not give what is intended when the axiom of foundation is not available. Hereditarily finite sets are those sets that are finite, whose members are finite, whose member's members are finite, and so on. It could happen that the words "and so on", just used to describe the hereditarily finite sets, mask an infinite process. Without the axiom of foundation an hereditarily finite set could have a countably infinite transitive closure. For this reason we should like to encourage Finsler's use of the term *totally finite* to describe the sets whose transitive closure is finite. In ZF, totally finite sets and hereditarily finite sets are the same. Additional details can be found in Booth [1990].

In *Totally Finite Sets* Finsler defines the product and sum of two sets. The definition is an extension of the product and sum of order types developed by Cantor and Hausdorff.

There have been other extensions of arithmetic to partial orderings, by Garrett Birkhoff in [1937] and [1942] for instance, but none of the definitions known to us agree with the ones here. Nevertheless, one feels that these definitions are quite natural ones.

The basic ideas for extending arithmetic to the totally finite sets is found in *Totally Finite Sets*. *On the Goldbach Conjecture* is a brief note that is easy to follow after reading only part of *Totally Finite Sets*. It has no special importance, but the idea of a set-theoretic Goldbach conjecture delighted us and we have made it the final article in this collection.

The knowledge of this set arithmetic was considerably advanced by Guerino Mazzola, [1969], [1972], [1973]. The first of these papers treats the arithmetic of sets in a somewhat different style than that of Finsler. The second paper, Mazzola [1972], solves a central problem left open in Finsler's work: Do the totally finite sets permit a unique factorization into primes? In [1973] Mazzola gave a characterization of the arithmetic on the hereditarily finite sets by using a projective property like those of free algebras.

There has not been an attempt to apply similar arithmetic concepts to the non-well-founded sets. The very problem of classifying non-well-founded sets is a rich field in itself. The following essay introduces the various combinatorial issues that arise in connection with non-well-founded sets. A reader who has not yet been stricken with an interest in non-well-founded sets can very well turn directly to *Totally Finite Sets*.

The Combinatorics of Non-well-founded Sets

Even though non-well-founded sets appeared in Mirimanoff [1917a] the most significant early treatment was that of Finsler [1926b]. Since then there have been studies of set theory without an axiom of foundation, but these have usually been relative consistency arguments that fell short of accepting non-well-founded sets themselves. The publication of Aczel [1988] marked a change in this respect: Aczel gave an *anti-foundation axiom* that replace the axiom of foundation. He also compared Finsler's axiom of identity to his anti-foundation axiom.

The subject is now sufficiently clear that one can at least survey the elementary facts. Open questions lie on all sides.

First, let us see the few examples of non-well-founded sets given by Finsler who did not attempt to systematically classify them. His examples were scattered about in [1926b] as illustrations of points in the development of set theory.

1. Finsler's Three Examples

Four graphs of non-well-founded sets appeared in Finsler [1926b], although, as will be explained, these examples give more than just four sets.

Example 1.1 (J-set) $J = \{J\}$.

The letter "J" is a graphically suggestive name for this set because the set, like the letter, hooks back on itself.

Example 1.2 (Fibonacci-Finsler sets) $A = \{A, B\}$, $B = \{A\}$.

Later we will use the letter K to designate these and other closely related sets. The name "Fibonacci is used because of the close connection between these sets and the sequence of Fibonacci numbers. The sets A and B may be compared respectively to the mature and immature states of Fibonacci rabbits. A mature pair preceeds both a mature and immature pair in the following month, whereas an immature pair yields only a mature pair. This single example 1.2 provides two sets that are defined simultaneously in a mutually dependent way.

The third example requires a preliminary definition. Regard the ordinal numbers as consisting of the set of lesser ordinals in the usual way: $1 = \{0\}$, $2 = \{0, 1\}$, $3 = \{0, 1, 2\}$ and so on. The related sequence of sets 0, $\{0\}$, $\{\{0\}\}$, $\{\{\{0\}\}\}$, ..., is also in need of a name. Let us devote the letter N to this use.

Definition 1.3 $N_0 = 0$, $N_{n+1} = \{N_n\}$.

Example 1.4 (Finsler's ladder) $M_n = \{N_n, M_{n+1}\}$.

In this third example there are infinitely many sets M_n defined simultaneously.

Finsler's fourth set diagram (Figure 4, p. 126 of this volume) is also infinite.

Besides these examples, Finsler provided two graphical devices for the study of non-well-founded sets: graphs and trees.

2. Graphs and Trees

The graph and the tree of a set are definable in terms of each other, and either of them can be used to recover the set itself. Both concepts depend on the transitive closure of a set. It is important, where non-well-founded sets are concerned, to distinguish between the transitive closure and the transitive hull.

Definition 2.1

(i) *A set S is said to be **essential in** R if S is an element of R, an element of an element of R, an element of an element of an element of R, and so on to any finite level of depth within R.*

(ii) *The collection of all sets S which are essential in R forms the **transitive closure**, $T(R)$, of R.*

(iii) *If we adjoin the set R itself to its transitive closure, we obtain the **transitive hull** $TH(R)$. In symbols $TH(R) = T(R) \cup \{R\}$.*

This distinction between transitive closure and transitive hull is not usually made in the set theoretic literature because in common circumstances only the transitive closure is required. But in the Finsler theory, unlike the usual set theories, sets may be members of themselves, so the distinction between the two notions becomes more important than usual.

Now we are ready for the concept of the graph of a set. Actually the concept of a graph is more general than that of a set and is logically identical to the concept of a relation. When the two notions, graph and relation are placed adjacent to each other, it is clear that they are the same.

Definition 2.2

(i) *A **graph** is a collection of points G, called nodes, along with a collection of arrows pointing from some nodes to others.*

(ii) *A **relation** is a set G of objects along with a collection of ordered pairs, ⟨ a, b ⟩, of objects of G.*

That the two concepts are identical is manifest. There is a slight pecularity in the mathematical literature however. A long tradition, going back to Euler, treats graphs whose edges run in both directions; that is, it treats only symmetric relations. This less general notion is often called a "graph" and the concept needed here is called an "undirected graph".

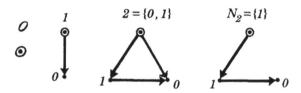

Figure 6

In Figure 6 the graphs of the first three finite ordinals 0, 1, 2 appear along with N_2 (definition 1.3). Turning to non-well-founded sets, the simplest one, example 1.1, is the J-set. Let us introduce a refinement to Finsler's graphical method and circle the node of the graph as shown at the right in Figure 7.

The reason for this circling is that graphs will generally have many nodes: We would like to display which one actually represents the the set that is the object of our attention.

As another illustration, example 1.2 above actually gives two Fibonacci sets. The graphs of these two sets must be kept distinct even though they are defined together as relations.

0　　　　*J*

Figure 7

To distinguish between them we indicate a root, as shown in Figure 8. When the node corresponding to A is the designated root, then we are interested in the set A, and likewise for the set B.

Combinatorics

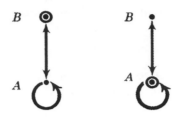

Figure 8

The distinguished node is said to be the *root*.

Definition 2.3 *A **rooted graph** is a graph on the set of nodes G for which one node, the root, is distinguished.*

Definition 2.4 *The **graph of a set** R is a rooted graph on the nodes G = TH(R) with R as its root. For any two sets A, B either essential in R or equal to R itself, that is A, B ∈ TH(R), place an arrow from A to B exactly if B ∈ A.*

Now, let us turn to the trees associated with rooted graphs.

Definition 2.5
(i) *A **path** in a graph is a sequence of nodes ⟨ D_1, D_2, ..., D_n ⟩ such that each node is connected to its successor in the sequence by an arrow in the graph.*
(ii) *A **rooted tree** is a rooted graph G with a root R such that each node can be reached from R by a unique path.*

In other words, a tree cannot be disconnected since nodes must be accessible from the root, nor can it have loops since nodes must be uniquely accessible.

It can easily be seen that each rooted, connected graph has an associated tree consisting of the paths through it. Before giving definitions it is best to examine some sample diagrams. Figure 9 shows a rooted graph. Figure 10 shows the tree of paths to which it gives rise. This process is called "unfolding" by Aczel [1988].

In the case of the ordinal numbers, the first three of which are shown in Figure 6, let us simplify the path tree somewhat by labeling its nodes only with the last term of the sequence that is properly attached there. In other words, in the tree for the ordinal number 3 = {0, 1, 2}, which

Figure 9

is given in Figure 11, there are four different nodes labeled "0". One of them is the path $\langle 3, 0 \rangle$ descending directly from 3 to its element 0. Another is $\langle 3, 1, 0 \rangle$ which goes from 3 to its element 1, and then on to 1's element 0. There are two more paths out of 3 and ending at 0, which go by way of 2.

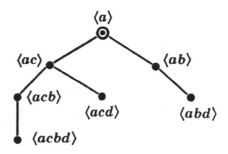

Figure 10

This simplified labeling, instead the full path label of figure 10, will be used here in drawing trees of sets. Literally, however, definition 2.5 identitifies nodes in the unfolding tree with a path.

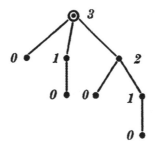

Figure 11

Definition 2.6 *The **unfolding tree** of a set R is the path tree of the graph of the set R.*

Roughly, speaking, Finsler [1926b] identified two sets if their graphs are isomorphic. In an alternate definition appearing in [1954] and described in [1964], he identitified sets whose graphs were compatible with each other. These two different methods of identifying sets are more fully described in section V of the introduction to the foundational section of this volume.

In [1988] Aczel independently used Finsler's second version, the more restrictive one, that all sets whose graphs are compatible should be identified. Aczel did not know that Finsler had already considered this criterion. In §22 of [1964] Finsler described this as a "higher" point of view for logical reasons, even though it produces fewer sets. We prefer Finsler's original, less restrictive criterion of set identity because we do not want to define sets out of existence without cause.

In [1960] Scott identified sets whose unfolding trees are isomorphic. This criterion of set identity is intermediate in strength between the most liberal criterion of Finsler's first version of axiom II and the most restrictive criterion used by Aczel and by Finsler in his second version of axiom II.

When we are concerned, as here, with combinatorial matters, it seems best to allow the widest variety of sets. If our interests are more purely logical, it may be better to restrict ourselves. The examples that follow shall illustrate this point of view.

3. *The Second Level*

Definition 3.1 *The level of a set S is the cardinality of $TH(S)$.*

No matter what criterion of set identity one may prefer, there are only two sets of the simplest kind, the sets of the first level. They are 0 and J.

The theorem that follows is a theorem of absolute set theory, which resembles naive set theory except that it is not naive; that is, vicious circles are not allowed though their existence is recognized. Circular definitions that can be consistently made are permitted. Technically speaking the proof is not completed because we have left open the criterion of set identity. We have described the criteria of Finsler, Scott, and Aczel but have not specified which is to be accepted. At the first level, however, these critera all have the same effect.

Theorem 3.2 *There are two sets of the first level.*

Proof. A single object can either be related to itself or to nothing at all. This produces two possible relations on one object. By any criterion of set identity, the graphs of distinct sets cannot be isomorphic. So we must have two distinct, possible sets, 0 and J. Each of these graphs has a unique root. Neither graph is dependent on any other since they have no non-empty subgraphs. It follows that both are consistently defined as sets, thus they actually exist as the sets 0 and J. This completes the proof. ■

A set of the second level has a graph of two nodes $\{a_1, a_2\}$. This graph has an adjacency matrix (e_{ij}) in which $e_{ij} = 1$ when $a_1 \in a_2$ and $e_{ij} = 0$ when $a_1 \notin a_2$. This gives a convenient enumeration of relations on two elements by the integers 0, 1, ..., 15 obtained by making (e_{ij}) correspond with

$$(\star) \qquad\qquad e_{11} 2^3 + e_{12} 2^2 + e_{21} 2 + e_{22}.$$

The permutation interchanging a_1 and a_2 induces internal automorphisms among the relations numbered 0, 6, 9, and 15, using the enumeration ($\star$). On the remaining relations it produces orbits of length two, pairing: 1 with 8, 2 with 4, 3 with 12, 5 with 10, 7 with 14, and 11 with 13. This leaves ten isomorphism types that could possibly be sets. The next theorem reduces the stock of suitable graphs to five.

Definition 3.3 *A graph is **extensional** if it does not contain two distinct nodes having the same immediate descendents.*

Definition 3.4 *A rooted graph is **accessible** if every node can be reached along a directed path from the root.*

Theorem 3.5 *The graph of a set is rigid, accessible, and extensional.*

Proof. The graph of a set is clearly rooted by the set itself, and every node is accessible from the root.

If the graph of R were not rigid, an automorphism would interchange two nodes P and Q. Since $TH(P)$ and $TH(Q)$ are subgraphs of $TH(R)$, they would be isomorphic. Even the least strict criterion of set identity will identify sets with isomorphic graphs. So P and Q are the same set. Thus they cannot arise as distinct nodes in the graph of R; no such automorphism can exist.

Finally, in order to see that the graph of a set is extensional, consider two nodes M and T having the same immediate descendents. These descendents must correspond to sets that are elements of M and T. So M and T must have the same elements. This means that the graphs of M and T are isomorphic. So by any criterion of set identity they are the same. Thus they could not give distinct nodes in the graph of R. This completes the theorem. ■

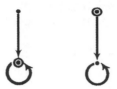

Figure 12

By these standards we may dismiss the non-rigid relations, numbered 0, 6, 9, and 15, in the enumeration (☆). The graphs 0, 1, 8, and 9 are disconnected, so they cannot be graphs of sets. One of the remaining six types of relation can be dismissed by the requirement of extensionality. The isomorphism type containing the relations numbered 5 and 10 in the enumeration can be given roots in two ways to produce the rooted graphs shown in Figure 12

The left hand graph in Figure 12 is connected as a graph but does not meet the stronger accessibility requirement that there be a path from the root to each node. The right hand graph is accessible but not extensional.

Four relations remain as possible graphs of sets. Three of them are accessible from only one of the two nodes. The remaining graph has two possible roots. These graphs are shown in Figure 13.

Figure 13

At this point the criterion of set identity becomes important. By the standards of Finsler [1926b], these graphs give five sets. By the standards of Finsler [1954] or Aczel [1988] they provide only two.

Definition 3.6

(i) *Let* **finsler(n)** *be the number of sets of level n such that sets are identified only if their graphs are isomorphic.*

(ii) *Let* **aczel(n)** *be the number of sets of level n such that sets are identified when it is consistent to do so.*

The previously given enumeration shows that *finsler*(2) ≤ 5. By checking that these five sets are well-defined, independent of the definition of any sets except those of level one, and have elements that are proper sets of a lower level, we have established the following.

Theorem 3.7 *finsler*(2) = 5.

The leftmost of the five sets of Figure 13, is the ordinal 1 = {0}. Let us give names to the others. The second from the left, the set most resembling 1, will be 1*. We may think of 1*, 2*, 3*, ... as an infinite list of pseudo-ordinals. The definition is given here for finite levels only but it continues into the transfinite.

Definition 3.8 $0^* = 0$ *and* $(n + 1)^* = \{0^*, 1^*, ..., n^*, (n + 1)^*\}$.

The third diagram from the left in figure 13 resembles the J-set; it will be called J_1.

Definition 3.9 $J_0 = J$ *and* $J_{n+1} = \{J_0, J_1, ..., J_n, J_{n+1}\}$.

Using the most strict standards of set identity, in which sets that can be consistently identified are actually taken to be equal, the sets J_n are not new sets. To see this, label both nodes in the diagram for J_1 (the third diagram from the left in Figure 13) with the letter "J". The resulting diagram does not contradict the definition of J. This test for strict set identity was introduced as an "anti-foundation axiom" by Aczel [1988].

Criterion 3.10 (Aczel's test) *There is a unique way to assign sets to a rooted, accessible graph so that the elements of a set are its immediate descendents in the graph.*

Since we can assign "J" to each node consistently, the two nodes must be the set J, otherwise there would be more than one way to label the graph with sets. The remaining pair of sets of level two, the right hand diagram in Figure 13, are also equal to J by Aczel's test. So there are only two Aczel sets of the second level, 1 and 1*, the other diagrams are merely elaborate characterizations of J.

Proposition 3.11 *aczel*(2) = 2.

Returning to Finsler's more liberal concept of set identity, the right hand pair of sets of the second level, Figure 13, are the

224   *Combinatorics*

Fibonacci sets given in example 1.2. These are also part of a series.
These sets, however, unlike n^* and J_n, have sets defined at level n
that are all distinct from those at previous levels.

There is a single Fibonacci set at the initial stage. Written, K_0^1, it
is the same set as the J set $J = \{J\}$. The superscript shows that it is of
the first level.

To continue the Fibonacci process, the next stage, the pair in
Figure 13, gives not one new set but two mutually dependent sets.
Call them K_0^2 and K_1^2. These two sets can be written with braces as:

$$K_0^2 = \{\, K_0^2, \, K_1^2 \,\}\,, \quad K_1^2 = \{K_0^2\}$$

These are the sets "A" and "B" of example 1.2.

At the next stage there are three mutually dependent sets:

$$K_0^3 = \{K_0^3, \, K_1^3, \, K_2^3\}\,, \quad K_1^3 = \{K_0^3\}, \quad K_2^3 = \{K_1^3\}$$

A complication here is that none of these sets are identical with
the ones defined at earlier levels, so superscripts have to be used to
distinguish them from the previous ones. The definition continues in
this way. At each stage all but one of the sets contain only a single
element.

Definition 3.12 *Let* $K_0^n = \{\, K_0^n, \, K_1^n, \, ..., \, K_{n-1}^n\}$ *and* $K_{i+1}^n = \{K_i^n\}$ *for*
$0 \leq i < n - 1$.

The graphs of the K-sets are most easily drawn with curved
arrows in Figure 14, where the diagram is given for the Fibonacci
sets of the third level.

It can easily be checked that the Finsler sets appearing in Figure
9 all have non-isomorphic trees. This implies that they are sets in the
sense of Scott too. In the next section a combinatorical tool, the layer
sequence, will be introduced; it can be used as a necessary condition
for trees to be isomorphic.

Figure 14

4. Layer Sequences

The layer sequence counts how many nodes lie at a given depth from the root of a tree. It provides a sufficient, but not necessary, condition for distinguishing between sets, and it connects pure set theory with enumerative combinatorics. The reader may skip this section entirely and continue with the survey of sets at the third level, beginning on page 228.

Definition 4.1

(i) *The root of a tree will be said to be at **layer** 0. If a node is immediately beneath a node of layer k, it will be said to be at **layer** k+1*.

(ii) *The **layer sequence** of a rooted tree is the sequence whose terms are the number of elements at each layer of the tree.*

(iii) *The **layer sequence** of a set R is the layer sequence of the unfolding tree of R.*

For example, in the tree of the ordinal number 3, shown in figure 9, one counts $\langle 1, 3, 3, 1 \rangle$ going down the layers of the tree. There is only one root at the top layer. Then there are three nodes at the first layer down, corresponding to the pathways out of $3 = \{0, 1, 2\}$ going to its three elements, and so on. It is sometimes helpful to think of all such sequences as being infinitely long so that the ordinal $3 = \{0, 1, 2\}$ gives the layer sequence $\langle 1, 3, 3, 1, 0, 0, 0, 0, ... \rangle$ rather than merely $\langle 1, 3, 3, 1 \rangle$.

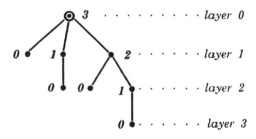

Figure 15

The reader may wish to draw the graph of $4 = \{0, 1, 2, 3\}$ and then draw the unfolding tree of this graph. It has 16 nodes given by paths through the graph. Finally, counting the number of objects in each layer, we obtain the layer sequence $\langle 1, 4, 6, 4, 1, 0, 0, ... \rangle$. This gives a suggestive idea of a connection between the finite ordinals and the binomial coefficients.

Definition 4.2
(i) *The **power series of a set** R is the series*

$$a_0 + a_1 x + a_2 x^2 + a_3 x^3 + \ldots = \Sigma\, a_n x^n$$

where $\langle a_0, a_1, a_2, \ldots \rangle$ *is the layer sequence of the set* R.

(ii) *The generating function corresponding to a set* R *is the algebraic fraction which is formally equivalent to the power series of the set* R.

We have seen that the ordinal $3 = \{0, 1, 2\}$ has the layer sequence $\langle 1, 3, 3, 1, 0, 0, 0, 0, \ldots \rangle$; so it has the power series: $1 + 3x + 3x^2 + x^3$. Since this power series terminates in finitely many steps, it is already in closed form and is its own generating function. We could, of course, write it as $(1 + x)^3$.

Theorem 4.3 *The generating function corresponding to the finite ordinal* n *is* $(1 + x)^n$.

Proof. The proof proceeds by induction on n. When $n = 0$, the ordinal in question is the empty set. Its graph is a single, isolated node. There is but one path in this rooted graph: the path which goes nowhere but stays at the root. Thus the unfolding tree is likewise a tree of one element. The layer sequence of such a trivial tree is $\langle 1, 0, 0, 0, \ldots \rangle$. So the power series of the empty set is $\langle 1 \rangle$. And this can certainly be expressed by the coefficients of $(1 + x)^0$.

Let us suppose that we know that the ordinal $m = \{0, 1, \ldots, m - 1\}$ has the generating function $(1 + x)^m$ for each $m < n$.

We now have to show that $n = \{0, 1, \ldots, n - 1\}$ also has a binomial generating function. The elements of n are $\{0, 1, \ldots, n - 1\}$, and these stand one layer beneath the root. So at level k in the tree of the ordinal n, we will find the same things that lie at level $k - 1$ of its elements. That is to say, at level k the ordinal n has

$$(*) \qquad \binom{n-1}{k-1} + \binom{n-2}{k-1} + \ldots + \binom{0}{k-1}$$

nodes in its unfolding tree. Of course, from a certain stage on these combination numbers will become 0. So, discarding the null terms, we could write

$$(**) \qquad \binom{n-1}{k-1} + \binom{n-2}{k-1} + \ldots + \binom{k-1}{k-1}.$$

Now this sum of binomial coefficients can be simplified by adding them together in pairs from the right using the identity

$$\binom{m}{i} + \binom{m}{i-1} = \binom{m+1}{i}$$

So, we may replace the last term of (**) by $\binom{k}{k}$, and then, using this identity, add together the last two terms of (**) obtaining

(***) $$\binom{n-1}{k-1} + \binom{n-2}{k-1} + ... + \binom{k+1}{k-1} + \binom{k+1}{k}.$$

Continuing this process the whole expression reduces to $\binom{n}{k}$.

This shows that the layer sequence of $n = \{0, 1, ..., n-1\}$ is also a binomial coefficient, completing the theorem. ∎

The same series of definitions – graph, tree, layer sequence, and generating function – can be applied to non-well-founded sets as well as to the finite ordinals. When this is done, however, the layer sequence is infinitely long.

The tree of the J-set, $J = \{J\}$, is an infinite chain given by endlessly circling paths around the sole loop of the graph in figure 2. Its layer sequence is plainly $\langle 1, 1, 1, 1, ...\rangle$; and the corresponding series is

$$1 + x + x^2 + x^3 + x^4 + ... = (1 + x)^{-1}.$$

The expression $(1 + x)^{-1}$, is the generating function for the layer sequence of the J set.

The set $J_1 = \{J, J_1\}$ can be seen from its tree to yield the layer sequence $\langle 1, 2, 3, 4, ...\rangle$. The next J set, $J_2 = \{J, J_1, J_2\}$, gives the sequence $\langle 1, 3, 6, 10, ...\rangle$, known as the *triangular numbers* since ancient times. The subsequent J set, J_3, gives the sequence $\langle 1, 4, 10, 20, ...\rangle$ which can be called, by analogy, the *tetrahedral numbers*. It is well known that these sequences are also binomial coefficients and form the diagonal sequences in Pascal's triangle.

Before determining the generating function of the J sets a lemma on formal power series is needed. It can be proved by comparing the terms in a formal product.

Lemma 4.4 *If $p(x)$ is the generating function of the power series $\Sigma a_n x^n$, then $(1 - x)^{-1} p(x)$ is the generating function for the series $\Sigma b_n x^n$, where $b_n = a_0 + a_1 + ... + a_n$.*

Theorem 4.5 *The generating function for the set J_n is $(1 - x)^{-n-1}$.*

Proof. It has already been checked for $n = 0$ that $(1 - x)^{-1}$ gives the layer sequence for the J set. Now, suppose that we have the theorem for $0, 1, ..., n - 1$.

It will prove useful to have notation for the layer sequences of J sets corresponding to our use of combination numbers in theorem 1. To this end, let $\langle J_n(0), J_n(1), J_n(2), ... \rangle$ be the layer sequence for the set J_n given in definition 3.9.

The induction hypothesis states that: $\Sigma J_i(k)x^k = (1 - x)^{-i-1}$, whenever $i < n$. The case $i = 0$ concerns the J-set $J = J_0$ which, as we have seen, has the layer sequence $(1 - x)^{-1}$.

To prove the theorem for $i = n$, one must show that the generating function for the set J_n is $(1 - x)^{-n-1}$. This factors into $(1 - x)^{-1} (1 - x)^{-n}$; so, by lemma 4.4, we need to show that

(1) $$J_n(k) = J_{n-1}(0) + J_{n-1}(1) + ... + J_{n-1}(k).$$

The notation here differs from that of lemma 4.4 in that the position of a term in the layer sequence is indicated in parentheses rather than as a subscript.

When $k = 0$, we have $J_i(k) = J_i(0) = 1$ for all i, since the trees of all J_i sets have a unique root. We may assume that (1) holds and by induction on k attempt to establish:

(2) $$J_n(k + 1) = J_{n-1}(k) + J_{n-1}(k) + ... + J_{n-1}(k + 1).$$

It is plain from the definition 3.9 that the nodes of layer $k+1$ in the unfolding tree of J_i come from layers k in the trees of $J_0, J_1, ..., J_i$. After all, these sets are the elements of J_i and form its upper layer. Thus we have that:

(3) $$J_i(k + 1) = J_0(k) + J_1(k) + ... + J_i(k).$$

In particular, we have that

(4) $$J_n(k + 1) = J_0(k) + J_1(k) + ... + J_{n-1}(k) + J_n(k).$$

Using (1), however, the last term can be replaced.

(5) $J_n(k + 1) = J_0(k) + J_1(k) + ... + J_{n-1}(k) + [J_{n-1}(0) + ... + J_{n-1}(k)].$

Now, applying (3) to the expression on the right which lies outside of brackets one obtains:

(6) $J_n(k + 1) = J_{n-1}(k + 1) + [J_{n-1}(0) + J_{n-1}(1) + \ldots + J_{n-1}(k)]$.

But except for the order of the terms, this is what was to be established in (2). This completes the theorem. ∎

The calculation of the generating function of the pseudo-ordinals, n^*, of definition 14 can be carried out in a similar way.

The Fibonacci sets are more removed from the finite ordinals than the J sets are, and they have more interesting layer sequences.

The unfolding tree of the pair K_0^2, K_1^2 is easily seen to be the tree appearing in the famous problem of Fibonacci's rabbits. To produce the literally correct Fibonacci sequence, $\langle 1, 1, 2, 3, 5, 8, 13, \cdots \rangle$, we must make K_1^2 the root as in Figure 11. When K_0^2 is the root we obtain an off-by-one Fibonacci sequence $\langle 1, 2, 3, 5, 8, 13, \cdots \rangle$ instead.

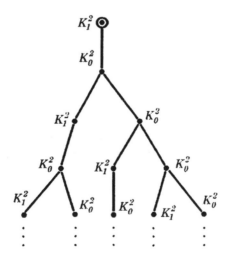

Figure 16

It is a well known fact that the generating function for the Fibonacci sequence, the layer sequence of K_1^2 is $(1 - x - x^2)^{-1}$. It is easy to obtain from this fact, by comparing terms of the respective power series, that the generating function for K_0^2 is given by the product $(1 + x)(1 - x - x^2)^{-1}$.

The next sets of the Fibonacci type are the triple K_0^3, K_1^3, K_2^3. Figure 17 shows the tree of K_1^3.

By counting the layers we intuitively form the sequence $\langle\ 1,\ 1,\ 1,\ 3,\ 5,\ 9,\ 17,\ \cdots\ \rangle$.This sequence has the property that each term after the third is the sum of the preceding three terms. It has been called the "tribonacci" sequence for this reason.

The very idea of forming the neologism "quatrobonacci" would probably horrify linguists, so one had better call the next sequence the Fibonacci sequence of the fourth level:

$$\langle\ 1,\ 1,\ 1,\ 1,\ 4,\ 7,\ 13,\ 25,\ 49,\ 94,\ \cdots\rangle.$$

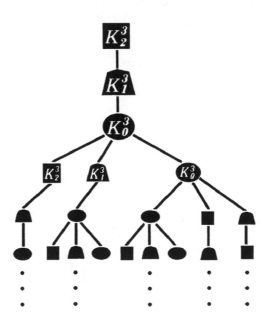

Figure 17

Definition 4.6 *The* **Fibonacci sequence of level** *n is the sequence* $\langle F_0^n,\ F_1^n,\ F_2^n,\ \cdots \rangle$ *where* $F_0^n = F_1^n = \cdots = F_n^n = 1$; *and*

$$F_{n+k}^n = F_k^n + F_{k+1}^n + \cdots + F_{k+n-1}^n$$

Theorem 4.7 *The layer sequence of the Fibonacci set* K_{n-1}^n *is the Fibonacci sequence of level n.*

This final theorem of the kind, theorem 4.7, will be stated without proof. The proof is a significantly more lengthy induction than the ones used previously here. This complexity arises because the Fibonacci sets defined at some given level do not coincide with those of earlier levels; this means that an inductive argument must proceed less directly. The proof will be given elsewhere.

A survey of all the possible Finsler sets shows that there are only five sets at the second level: 1, 1^*, J_1, K_0^2 and K_1^2.

All five of these sets have been shown to be parts of more general collections of finite Finsler sets, all of which have ties to familiar objects of combinatorial theory.

Among the sets of the third level, those whose transitive closure has three elements, there are many non-well-founded sets other than the J sets, pseudo-ordinals, and Fibonacci sets. Among these new kinds of sets are some instructive examples that do not appear at the first two levels.

5. The Third Level

There are several new phenomena that arise for the first time at the third level of sets. The most striking is an example of a set theoretic equation that has no unique solution. This example has several interesting properties: We might even call them paradoxical.

Consider the Fibonacci set of the second level $K_0^2 = \{K_0^2, K_1^2\}$. These are the sets of example 1.2. Finsler called them by generic names "A" and "B".

Next, consider the following equation of set theory where X is a variable: $X = \{X, K_1^2\}$. Of course, $X = K_0^2$ is a solution to this equation. But there is another possible solution which shall be written "G", though the name is merely a temporary one.

Example 5.1 $G = \{G, K_0^2\}$ and $G \neq K_0^2$.

At first one can only say that this defines a class. It surely satisfies Finsler's axiom I. If axiom II were to require that $G = K_0^2$, then example 5.1 taken as a definition would be inconsistent. So it must be shown that the two sets can be consistently kept distinct.

To see that they are distinct, form the transitive hulls of both sets. Since they are each essential in themselves, the transitive hulls and transitive closures are the same. The set K_0^2 is transitive, $TH(K_0^2) = TC(K_0^2) = K_0^2$, and $TH(G) = TC(G) = \{G, K_0^2, K_1^2\}$. Thus, the graphs of these two sets can consistently be kept distinct. If we assume that G is different from K_0^2 and K_1^2, there is nothing that forces us to retract the assumption when we examine the structure of these sets. The set G is of the third level, its transitive closure has three elements: The set K_0^2 as we have seen, is of the second level. It is easy to see that the equation $X = \{X,\ K_1^2\}$ can have no solutions other than these two, G and K_0^2.

According to the most strict standards of set identity, that two sets are to be regarded as the same when it is consistent to do so, we could not have $G \neq K_0^2$. This is plain from the expression of the sets in bracket notation. This peculiar fact, that set equations may have more than one solution, is probably the reason that Finsler adopted a stricter notion of set identity in his later papers.

Another phenomenon that first appears at the third level concerns subgraphs of the graphs of a set. In Figure 18 there is a graph of three elements showing a specially marked subregion.

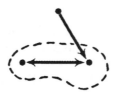

Figure 18

In this graph the enclosed region is self contained, that is, arrows enter it but do not leave. The nodes within this region must also stand independently as sets if the entire graph represents a set. In Figure 18 this does not happen because the enclosed region, or *district* as we shall call it, possesses a automorphism.

Definition 5.2 *A **district** is a subgraph in which nodes of the district are related only to other nodes of the district by the unrestricted graph.*

In other words, arrows which originate within a district must remain there.

Definition 5.3 *A graph is **firm** if no district is isomorphic to a different district and no district has an automorphism other than the identity.*

Using these definitions theorem 11 can now be strengthened to require that districts within a graph properly define sets too.

Theorem 5.4 *The graph of a set is firm.*

Proof. Suppose there were an isomorphism between the districts D and E taking $d \in D$ to $e \in E$. Since the transitive hull of d will lie entirely within D and that of e within E, it must be that the graphs of d and e are isomorphic and therefore they are the same set. So the isomorphism would have to be the identity. ∎

As a result of Theorem 5.4 the graph of Figure 18 is not a set.

Example 5.5 $A = \{B\}$, $B = \{A, C\}$, $C = \{A, B\}$.

The graph given by this example is firm and extentional. Either A, B, or C can be a root and still have accessibility. Finsler's axioms can be used to show that these are all bona fide sets. To see that Finsler's axiom II holds, one must check that the sets are distinct and different from sets of lower levels.

The set A has a tree isomorphic to the full Fibonacci tree of K_1^2, Figure 16. So by the standards of Scott [1960] the sets would be identified to yield $K_0^2 = B = C$ and $A = K_1^2$.

These circular set definitions resemble situations in automata theory or in the theory of feedback, additional details can be found in Booth [1991]. In any case, it is clear that A, B and C could consistently be collapsed to the J-set, $J = \{J\}$. Similar sets can be defined at any finite level, as in example 5.6.

Example 5.6 $A_0 = \{A_1\}$, $A_1 = \{A_0, A_2\}$, $A_2 = \{A_0, A_3\}$, $\cdots$, $A_n = \{A_0, A_1\}$.

Example 5.7 $P = \{Q, 0\}$, $Q = \{P, Q, 0\}$.

In example 5.7 either P or Q can serve as a root. The set P has the same layer sequence, $\langle 1, 2, 3, 5, 13, \cdots \rangle$ as K_0^2 but has a tree that is not isomorphic to the tree of K_0^2. This example is also instructive because it produces two sets, P and Q, that are very closely related to the sets K_1^2 and K_0^2 respectively, yet the former can be consistently collapsed to the set 1^*, while the latter collapses to J. To see this one can employ the criterion 3.10.

To survey the remaining sets of the third level, replace the relations among three objects by their adjacency matrix, a 3×3 matrix of 0's and 1's, as was done in section 3 for the second level. This gives $2^9 = 512$ relations. As before, the matrix may be read as a binary integer; giving a convenient enumeration of the relations: 0, 1, ⋯, 511.   An argument using Burnside's theorem, which is described in general in Davis [1943], shows that these relations fall into 104 isomorphism types. The isomorphism types of firm, extensional, root accessible graphs at the third level all turn out to be Finsler sets.

Estimating the number of Finsler sets of an arbitrary level is complicated by the fact that various numbers of sets are defined by each of the suitable graphs.

For example, some graphs define new sets at every node. At the third level there are 14 such graphs, shown in Figure 19.

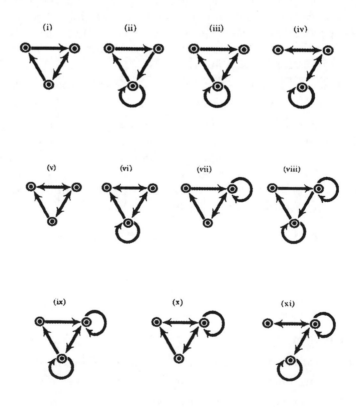

Figure 19

These 3×14 = 42 sets would all collapse to the *J*-set were we to adopt the most conservative standards of acceptance for non-well-founded sets, the criterion 3.10. They sets can be given in set theoretic notation too.

(i)	$A = \{B\}$	$B = \{C\}$	$C = \{A, B\}$
(ii)	$A = \{B\}$	$B = \{C\}$	$C = \{A, C\}$
(iii)	$A = \{B\}$	$B = \{C\}$	$C = \{A, B, C\}$
(iv)	$A = \{B\}$	$B = \{A, C\}$	$C = \{B, C\}$
(v)	$A = \{B\}$	$B = \{A, C\}$	$C = \{A, B\}$
(vi)	$A = \{B\}$	$B = \{A, C\}$	$C = \{A, B, C\}$
(vii)	$A = \{B\}$	$B = \{B, C\}$	$C = \{A, B\}$
(viii)	$A = \{B\}$	$B = \{B, C\}$	$C = \{A, C\}$
(ix)	$A = \{B\}$	$B = \{B, C\}$	$C = \{A, B, C\}$
(x)	$A = \{B\}$	$B = \{A, B, C\}$	$C = \{A, B\}$
(xi)	$A = \{B\}$	$B = \{A, B, C\}$	$C = \{B, C\}$
(xii)	$A = \{B\}$	$B = \{A, B, C\}$	$C = \{A, C\}$
(xiii)	$A = \{B, C\}$	$B = \{A, B\}$	$C = \{A, B, C\}$
(xiv)	$A = \{A, B\}$	$B = \{B, C\}$	$C = \{A, B, C\}$

Example (v) in this list has been given in Aczel [1988, 54].

There are also graphs which define only two new sets – the third node must be either 0 or *J*. These are given as graphs in Figure 20, and also listed using braces.

(i$_a$)	$A = \{B\}$	$B = \{A, 0\}$
(i$_b$)	$A = \{B\}$	$B = \{A, J\}$
(ii$_a$)	$A = \{B\}$	$B = \{A, B, 0\}$
(ii$_b$)	$A = \{B\}$	$B = \{A, B, J\}$
(iii$_a$)	$A = \{B, 0\}$	$B = \{A, B\}$
(iii$_b$)	$A = \{B, J\}$	$B = \{A, B\}$
(iv$_a$)	$A = \{B, 0\}$	$B = \{A, B, 0\}$

$$(\text{iv}_b) \qquad A = \{B, J\} \qquad\qquad B = \{A, B, J\}$$
$$(\text{v}_a) \qquad A = \{A, B\} \qquad\qquad B = \{A, B, 0\}$$
$$(\text{v}_b) \qquad A = \{A, B\} \qquad\qquad B = \{A, B, J\}$$

The examples occur in pairs: One member of the pair has a J-set where its mate has an empty set. If we were to identify sets when it is consistent to do so, then the left hand graph in each pair represents the set 1^* while the right hand set represents J. Of course, in the Finsler theory sets have their own independent existence whenever that is consistent, so all of these diagrams define two new sets (one at each of the encircled nodes).

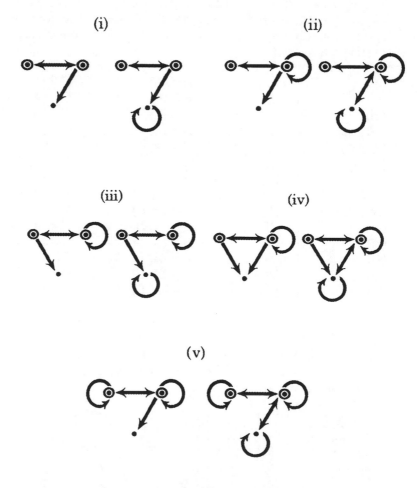

Figure 20

These ten graphs provide 20 sets, two for each diagram. The set given on the right of (iii) in figure 20 – it is (iii b) in the table on the previous page – was given as an example in Aczel [1988, p. 55].

Finally, there are 16 more sets which arise from graphs of three nodes, two of which correspond to sets of the first or second level. Six of these, shown in Figure 21, are variations on nested singleton sets made by attaching self-membership loops. The loops can be placed arbitrarily except that the two nodes at the bottom must never resemble the non-extensional graph of on the right of Figure 7.

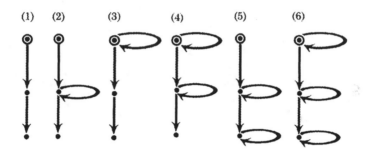

Figure 21

In general, there would be $3 \cdot 2^{n-2}$ sets of this kind at level n.

At last, we have obtained some sets of the third level that remain sets even under strict conditions of identity. The last two in figure 20, numbers 5 and 6, can be consistently labeled "J" at every node: Thus they fail to be sets by Aczel's criterion 3.10. The first four graphs, however, define sets by any standards of set identity.

The sets in figure 21 are listed in brackets below:

(1)	$\{1\} = \{\{0\}\}$	(4)	$A = \{A, 1^*\}$
(2)	$\{1^*\}$	(5)	$\{J_1\}$
(3)	$A = \{A, 1\}$	(6)	$A = \{A, J_1\}$

There are other diagrams, besides those of Figure 21, that define a single Finsler set of the third level. Just as loops were inserted into the diagrams of Figure 21, we can also put loops onto the graph of an ordinal number. These generalized ordinals resemble ordinal numbers: They are all transitive sets whose members are transitive sets, but they can be, of course, non-well-founded. We may call them *mirage ordinals*. There are four of them at the third level. Continuing with the numbering of the sets begun in Figure 21, these mirage ordinals – a real ordinal is among them – will be numbered 7-10.

Two of these ordinal mirages, 2^* and J_2, have been defined in section 3. One might have expected as many as eight mirages, including the real ordinal $2 = \{0, 1\}$, but the requirement of extensionality leaves only these four. For example, $\{0, 1^*\}$ is 1^* and not a set of the third level; likewise $\{J, J_1\}$ is J_1.

The following table and graphs give $2 = \{0, 1\}$ and its mirages.

(7)	$2 = \{0, 1\}$	(9)	$2^* = \{0, 1^*\}$
(8)	$A = \{A, 0, 1\}$	(10)	$J_2 = \{J, J_1\}$

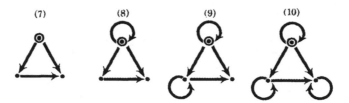

Figure 22

The last example in this series, 10 in Figure 22, can be labeled "J" at every node, so by the strict criterion 3.10 it is merely an elaborate definition of the J-set. The other graphs, however, have a node with no descendents which must be labeled as the empty set. These graphs define new sets by any standards.

The remaining sets of the third level are given in the table below and shown in figure 22.

(11)	$\{K_1^2\}$	(14)	$A = \{A, K_0^2, K_1^2\}$
(12)	$A = \{A, K_0^2\}$	(15)	$A = \{A, 0, 1^*\}$
(13)	$A = \{A, K_1^2\}$	(16)	$A = \{A, J, J_1\}$

Four of these examples have Fibonacci sets within them. Example 11 in Figure 23 was previously given as example 5.1.

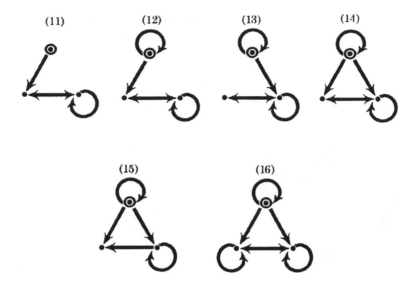

Figure 23

These 16 sets, the 20 that are in pairs in figure 19, and the 42 in figure 19 give 78 Finsler sets.

Proposition 5.8 *finsler*(3) = 78.

This value was given in Booth [1990] although there are several errors there in the listing that have been corrected here but did not effect the total. Only nine of these sets satisfy the most strict criterion of set identity, 3.10.

6. *The Countable Level*

Two of Finsler's original examples, Figures 4 and 5 of this volume, involved sets at a countable level. The graph of Figure 4 defines sets that meet the strict standards of set identity of criterion 3.10. The example is particularly instructive as giving non-well-founded sets that are nevertheless circle-free.

If we loosen the standards of set identity, then, as we would expect from the finite levels, many more sets appear. A particularly elementary example is shown in Figure 24.

Figure 24

By strict standards of set identity, $E_0 = E_1 = \ldots = J$, since the nodes of the graph can be labeled "J" without violating the facts of set membership.

We can change various of the double arrows in Figure 19 to single arrows in order to obtain uncountably many hereditarily finite sets. We can also obtain uncountably many hereditarily finite sets that meet the stricter conditions of 3.10.

There are a number of other instructive examples of non-well-founded sets at the countable level; but there is not yet a general theory concerning them.

In the past combinatorial set theory has treated problems concerning the relations between various classes of subsets of a şet. When we look beyond the familiar well-founded sets towards the non-well-founded sets, a new variety of combinatorial set theory meets our gaze. The examples given here indicate a close connection between the structure of non-well-founded sets and parts of traditional combinatorial theory: graph theory and enumeration theory.

First published as: "Totalendliche Mengen" (Dedicated to B. L. van der Waerden in honor of his 60th birthday.), *Vierteljahresschrift der Naturforschenden Gesellschaft in Zürich* **108** (1963), 141–152 (MR **37**, 6189).

Totally Finite Sets

§1. Sets of Finite Depth

The *sets* which are considered here are *pure* sets, i. e. their *elements* too are pure sets.

The *empty set*, which possesses *no* element, and the *unit set* which contains *only the empty set* are *examples* of such sets.

The elements of a set m, the elements of these elements etc., form the *transitive closure* of m. These sets will be called *essential in* m. If a set a is essential in b and b in c, then a is also essential in c.

If the transition from a set m to its elements, then to the elements of these elements, etc. can be made only *finitely* often, then this set m is said to be of *finite depth*. The sets essential in a set of finite depth are likewise of finite depth.

A set of finite depth is never essential in itself, otherwise this transition would not terminate. If a set b is essential in a set a which is of finite depth, then a is not essential in b, for otherwise a would be essential in itself.

If the transition from a set m to its elements, then to the elements of the elements, etc. can be carried out exactly s times, then s is said to be the *depth of the set* m. Thus the empty set has depth 0: The unit set is of depth 1.

A set of finite depth, together with the sets essential in it, has a definite *structure* which involves the sets essential in it. If the sets a and b have the same structure, i. e., if these sets together with the sets essential in them can be mapped onto one another in a one to one, element-preserving fashion, then they are *identical*, but otherwise they are *different*. The unit set is different from the empty set, because it possesses one element, whereas the empty set possesses no elements. There is, however, only *one* empty set and only *one* unit set.

The *elements* of a set must always be *different* from one another.

A set is said to be *finite* if it possesses only finitely many elements; the number of these elements is called the *cardinal number* of the set.

A set is said to be *totally finite* if it itself and all sets essential in it are finite, and if in addition it is of finite depth.

It will be shown that: *Every set of finite depth is totally finite.*

The elements of a set m form the *first depth* beneath m, the elements of these elements the *second depth*, etc. A fixed set, e.g. the empty set, can belong to different depths of m and can also appear more than once at a given depth. But it is to be counted only *once*. The *last depth* down from m necessarily consists only of the empty set, a finite set.

If the k-th depth from m consists of only finitely many finite sets, then the same holds also at the depth $(k - 1)$. The proposition follows from this by induction.

The totally finite sets together form a *cluster*, which we shall analyze.

§2. Sets as Numbers

The number 0 can be represented by means of the empty set: the number 1 by the unit set.

Consider the *natural numbers* 1, 2, 3, ... ; each successive one can be represented by the set which contains the *immediately* preceeding one as its sole element, thus: $2 = \{1\}$, $3 = \{2\}$, etc. We have also $1 = \{0\}$, but 0 is not a natural number. One might call 0 a *vanishing* number.

The natural numbers are then represented by totally finite sets, of which each possesses exactly *one* element; hence they are sets with cardinal number 1. The depth of a natural number coincides with the number itself.

The *ordinal numbers* **0**, **1**, **2**, **3**, ... can be defined so that each ordinal number represents the set of *all* previous ordinal numbers, thus $0 = \{\ \}$, $1 = \{0\}$, as before, but $2 = \{0, 1\}$, $3 = \{0, 1, 2\}$, etc.

The sequence of ordinal numbers can be continued into the transfinite. The *finite* ordinal numbers are all totally finite sets: Their cardinality and depth both coincide with the number itself.

Let us now regard *all totally finite* sets as *generalized numbers*, or simply as *numbers*, since we will not refer to numbers of any other kind. *Unlike* the natural numbers the generalized numbers may be of any finite power. Their depth is also finite. Now it will be shown that addition and multiplication can be introduced into the generalized numbers with certain modifications, in such a way that these operations act correspondingly on the depth numbers.

§3. The Order of the Numbers

The cardinality of an arbitrary number z will be denoted by z^*, the depth as $|z|$. For the natural number n: $n^*=1$ and $|n|=n$.

In order to put the numbers, that is to say the totally finite sets, into a simply ordered sequence, one could order them in the first instance by increasing depth, those of the same depth according to their cardinality, and those of equal power according to their elements, which we may assume ordered already.

To survey the *number* of sets at a fixed depth, one can first order *all the numbers up to a fixed depth s* in a definite way. Placing these orderings one behind the other for increasing s; then delete the numbers of smaller depth which have already been listed.

The *provisional ordering* of all numbers z such that $|z| \leq s$ will be defined as follows: Suppose that the ordering of the numbers y with $|y| < s$ is already known. The numbers z such that $|z| \leq s$ shall then be ordered according to their cardinality; those with equal cardinality are ordered lexicographically with respect to the known ordering for the elements of y.

In this way one obtains provisionally ordered sequences of numbers at each depth s: for $s = 0$, the number 0. For s = 1, it is the sequence 0, 1. For $s=2$, it is the sequence 0, 1, 2, **2**. For $s = 3$ one obtains, when the cardinal number $z^* = 0$, the number 0; then there follows, with $z^* = 1$, the numbers 1, 2, 3, {2}; when $z^* = 2$, the numbers **2**, {0, 2}, {0, **2**}, {1, 2}, {1, **2**}, {2, **2**}; when $z^* = 3$, the numbers {0, 1, 2}, **3**, {0, 2, **2**}, {1, 2, **2**}; and finally, when $z^* = 4$, the number {0, 1, 2, **2**}.

The number of numbers z such that $|z| \leq s$ is thus given by $2^0 = 1$, $2^1 = 2$, $2^2 = 4$, $2^4 = 16$ for $s = 0, 1, 2, 3$ respectively.

If in general the number of numbers y such that $|y| < s$ is equal to n, then for the numbers z such that $|z| \leq s$ one obtains $\binom{n}{m}$ of them having cardinal number $z^*=m$. Thus there are 2^n numbers altogether. For $s = 4$ there are $2^{16} = 65536$, for $s = 5$, 2^{65536}, etc.

The number of numbers for which the depth number is exactly equal to s amounts therefore to 1, 1, 2, 12, 65520, $2^{65536} - 65536$, etc., for $s = 0, 1, 2, 3, 4, 5$, etc. respectively.

One obtains the *final ordering* of the numbers by deleting all numbers satisfying $|z| < s$ from the provisional ordering of the numbers z such that $|z| \leq s$ and placing the sequences so obtained for increasing s one behind the other.

The numbers such that $|z| = s$ remain ordered according to their cardinal number. For numbers with the same cardinal number, however, there can result certain transpositions when the final ordering is compared with the provisional ordering.

For the numbers z with $|z| \leq 3$ the final ordering is different from the provisional ordering only in that the number **2** moves from the sixth to the fourth place, and thus comes to stand between the numbers 2 and 3, while the numbers 3 and {**2**} move to fifth and sixth places respectively. A comparison of the provisional and final ordering shows, for example, that for those z satisfying $|z| = 4$ and $z^* = 2$, the provisional rule is not fulfilled in the final ordering. The number {3, {**2**}} occurs before the number {3, **2**}, whereas according to the first rule the reverse would have to hold.

§4. *The Figure of a Number*

To each number one can associate a *figure* which is composed of *points* and *arrows*.

For each number itself and each number essential in it there is a point of the figure. From the point, a, corresponding to the number a, there proceeds an arrow to the point b, if b is an element of the number a. One can also call the points of such a figure the numbers of the figure.

The arrows can be replaced by *line segments* if one specifies that these shall always be orientated "from above to below". An orientation of the diagram is then required; the line segments can be inclined to this primary direction but are not allowed to be perpendicular to it.

The "highest" point of the figure of a number z represents z itself: The "lowest" point represents the empty set, that is the number 0. The empty set, zero, is essential in every number which is different from 0.

Each figure of a number is *connected*; by starting from the highest point one can reach every point of the figure by a connected sequence of arrows.

For the numbers z such that $|z| \leq 3$ one obtains the figures for the provisional ordering that are shown on the following page:

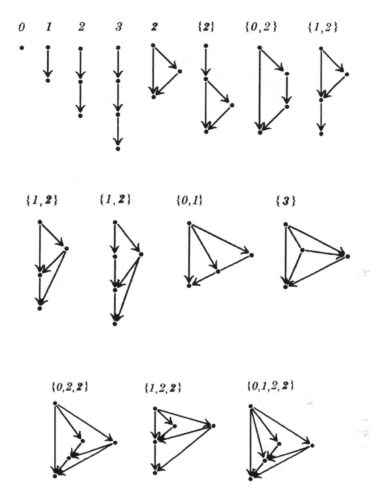

Figure 24

In order to make the different depths of a number z clear, it can be useful to employ another representation in which the elements of z are represented by points in an horizontal row, the elements of these elements in the second row etc. Again each number is connected to its elements by means of an arrow (or a line segment). In these diagrams a number can appear in more than one depth in a diagram, so they are no longer uniquely associated with a point.

The last number in figure 24, with $|z| = 3$, that is the number $\{0, 1, 2, \mathbf{2}\}$, has the following form:

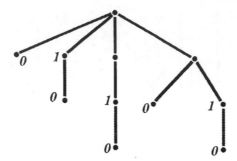

Figure 25

The previous kind of figure can be thought of as *flowers* and the last as the corresponding *umbels* of the clusters which have already been mentioned.

§5. Addition

The figure of the *sum a+b* of two numbers a and b is obtained by replacing the point 0 in the figure for a by the point b of the figure for b, that is by "attaching" the two diagrams so that the "lowest" point of the figure for a coincides with the "topmost" point of the figure for b. This defines the sum $a + b$.

The following rules result:

Addition is associative, i. e., $a + (b + c) = (a + b) + c$. This is directly evident.

Addition is not always commutative; for example, $1 + 2 = \{2\}$ but $2 + 1 = \{1, 2\}$.

We always have $0 + a = a + 0 = a$.

From $a + b = a$ it follows that $b = 0$; and from $a + b = b$ follows $a = 0$.

The depth of a number z is equal to the greatest number of arrows forming a directed path from z to the point 0. From this follows: *When the numbers are added of numbers their depth numbers are also added*, i. e. $|a + b| = |a| + |b|$.

For the natural numbers this addition coincides with the usual addition, and therefore $m + n = n + m$ holds for natural numbers m and n.

The cardinal number of a sum is equal to the cardinal number of the first non-vanishing term of the sum, i. e. $(a + b)^* = a^*$ for $a \neq 0$.

A sum of at least two non-vanishing ordinal numbers is not an ordinal number, since otherwise the element number of the first term of the sum would has to increase itself.

Since the empty set is essential in every non-vanishing number, the following rule holds: *If $a \neq 0$, then one obtains the sum $a + b$ by first adding the number b to each element of a and then forming the set of these numbers.*

Together with the rule $0 + b = b$ this provides an *inductive definition of addition* since the elements of a possess a depth less than that of the set a itself.

In particular $1 + b = \{0\} + b = \{b\}$ and, for example, $2 + 1 = \{0 + 1\} + 1 = \{1, 2\}$.

§6. Mononumbers

A number which can be represented as the sum of two non-vanishing numbers is said to be "additively composite" or, more briefly, *decomposable*.

The non-decomposable, that is to say the additively non-decomposable numbers, with the exception of 0, are called *mononumbers*.

The number 0 is not a mononumber, just as the number 1 is not a prime number.

The number 1 is a mononumber; the rest of the *natural* numbers are decomposable.

In order to determine all mononumbers it is simplest to delete 0 and the decomposable numbers from the sequence of all numbers as in the "sieve of Eratosthenes".

Among the numbers which have depth 2, the number $2 = 1 + 1$ is decomposable; the number $2 = \{0, 1\}$ is a mononumber.

When numbers are added their depths are also added, so one easily finds that the numbers $1 + 2 = 3$, $1 + 2 = \{2\}$, $2 + 1 = \{1, 2\}$ are the decomposable numbers of depth number 3; consequently there are 9 mononumbers of depth number 3.

Amongst the 65520 numbers of depth 4 one finds 12 of the form $1 + x$, in addition to these there are 10 more of the form $x + 1$, and finally the number $2 + 2 = \{2, \{2\}\}$. Altogether there are 23 decomposable numbers; accordingly there remain 65497 mono-numbers of depth number 4.

Among the numbers of depth number 5 there are 131047 decomposable numbers: namely 65520 of the form $1 + y$; in addition to these, 65508 more of the form $y + 1$; there are 10 more of the form $2 + x$; and an additional 9 of the form $x + 2$; where in each case y has depth 4 and x has depth 3. Consequently there remain $2^{65536} - 196583$ mononumbers of depth 5.

§7. The Decomposition of Numbers into Mononumbers

A number k essential in the number z is said to be a *knot* of z, if the figure of z splits up with the removal of the point k so that it is disconnected.

The figure of z will then split into two parts. The second part contains the numbers essential in k including 0; together with k they yield the figure for k. The first part contains the numbers which can still be reached by directed paths from z. The number k is essential in them, since each of these outgoing sequences of arrows, leading ultimately to 0, must pass through the point k. If one replaces the point k by the point 0 to make the knot into the "lowest" point of the figure, then one obtains the figure of a number a which satisfies $z = a + k$. Thus the number z is decomposable.

Since conversely a decomposable number z, where $z = a + b$ and $a \neq 0$, $b \neq 0$, possesses the knot b it follows that: *The non-vanishing numbers which do not possess knots are mononumbers.*

One easily finds that: If k is a knot of a, then $k + b$ is a knot of $a + b$; if l is a knot of b, then l is also a knot of $a + b$. Thus, for $a \neq 0$ and $b \neq 0$, the number $z = a + b$ possesses exactly one knot more than the numbers a and b together.

If k and l are distinct knots of z, then either k is essential in l or l is essential in k depending on whether l belongs to the first or to the second part of the decomposition of z at k. If k is essential in l, and l in m, then k is also essential in m. Thus the sequence of knots is linearly ordered.

One can "resolve" the knots of a number z, in an arbitrary order, replacing them by sums; this yields a representation of z as the sum of $n + 1$ mononumbers, when the number z has n knots. This representation is unique with respect to the order of the terms of the sum. The diagrams of the summands lie before, between, and after the knots of z.

If one also says that the number 0 is representable as the sum of 0 mononumbers, and every mononumber as the sum of one mononumber, then it follows that: *Each number is uniquely representable as a sum of mononumbers.*

From the unique decomposition of numbers into mononumbers there follows a lemma, which will be useful later: *If $a + b = c + d$, then $|a| = |c|$ and $|b| = |d|$ imply that $a = c$ and $b = d$.*

The depth numbers of the mononumbers are always at least 1 and are themselves added when mononumbers are added. The first mononumbers in the complete decomposition of the number $z = a + b = c + d$ produce a total depth of $|a| = |c|$. So these numbers are the same in both sums; this means, however, that $a = c$ and consequently $b = d$. The same result follows if one considers the sum of the last mononumbers with total depth number $|b| = |d|$.

§8. Multiplication

The figure of the *product ab* of two numbers a and b is obtained by replacing each arrow in the diagram of a by the diagram of b. The product ab is defined in this way. This product of the numbers is not the same as the usual Cartesian product in set theory.

This yields the following rules:

Multiplication is associative: $a(bc)=(ab)c$. This is directly evident.

Multiplication is not always commutative. For example, $2 \cdot 2 = 2 + 2 = \{2, \{2\}\}$, but $2 \cdot 2 = \{1, 3\}$.

It always holds that $0 \cdot a = a \cdot 0$ and $1 \cdot a = a \cdot 1 = a$.

From $ab = 0$ it follows that $a = 0$ or $b = 0$; from $ab = a$ it follows that $b = 1$; and from $ab = b$ it follows that $a = 1$.

These results can either be obtained directly or from the following rule: *When numbers are multiplied their depth numbers and cardinal numbers are multiplied as well*: $|a| \, |b| = |ab|$ and $a^*b^*=(ab)^*$.

The "longest" path from a to 0 in the figure of a contains $|a|$ arrows, the longest path from b to 0 in the figure for b contains $|b|$ arrows. Thus if one replaces each arrow in the figure of a by the figure of b, one obtains $|a| \, |b|$ as the length of the longest path from ab to 0 in the figure of ab.

In the figure of a, a^* arrows proceed from the point a, and in the figure of b, b^* arrows from the point b, thus there are a^*b^* arrows from the root of ab.

If n is a natural number, then $na = a + a + ... + a$ is equal to the sum of n terms each equal to a.

The figure for n consists of n arrows following one after the other which are then each replaced by the figure of a, yielding the figure of the sum $a + a + ... + a$.

In particular, the product mn of the natural numbers m and n has the usual meaning; furthermore $mn = nm$.

The *distributive law* in the form $(a + b)c = ac + bc$ holds in general, as can be seen directly.

The left distributive law however can fail, as in: $2(1 + 1) = 2 \cdot 2 \neq 2 \cdot 1 + 2 \cdot 1 = 2 \cdot 2$.

A product of ordinal numbers is in general not an ordinal number: for example, $2 \cdot 2 = \{0, 1, 2, \{2\}\} \neq 4 = \{0, 1, 2, 3\}$.

§9. Commutative Sums

The relation $a + b = b + a$ holds for *mononumbers* only if $a = b$. This follows directly from the uniqueness of the additive decomposition of $a + b$, since the first term of the decomposition is uniquely determined.

For arbitrary numbers $a \neq 0$ and $b \neq 0$ the following proposition holds: If $a + b = b + a$ then $a = mc$ and $b = nc$ where m and n are natural numbers.

For the proof let the numbers a and b be represented as sums of mononumbers: $a = d_1 + d_2 + ... + d_k$, $b = e_1 + e_2 + ... + e_l$.

From the uniqueness of the decomposition of the number:

$$z = a + b = b + a$$

into mononumbers, one has for $k > l$ that $d_1 = e_1$, $d_2 = e_2$, ..., $d_l = e_l$. Now put $d_{l+1} + ... + d_k = b_1$ so that $a = b + b_1$ and $z = b + b_1 + b = b + b + b_1$. Thus $b_1 + b = b + b_1$.

Now $z_1 = b_1 + b = b + b_1$ can be treated similarly. Whereas $z = a + b$ splits into $k + l$ monoterms in its sum, $z = b + b_1$ contains only k such terms.

If one now puts $k = l + l_1$, then one can write $l = l_1 + l_2$ and $b = b_1 + b_2$ for $l > l_1$, where $z = b_1 + b_1 + b_2 = b_1 + b_2 + b_1$. This gives:

$b_1 + b_2 = b_2 + b_1 = z_2$. For $l_1 > l$, however, one places $l_1 = l + l_2$ and $b_1 = b + b_2$ where $z_1 = b + b_2 + b = b + b + b_2$, yielding $b_2 + b = b + b_2 = z_2$. In either case z_2 contains fewer terms than z_1. One can continue in this way. Since the number of terms of the sum cannot decrease infinitely one comes to

$$z_r = b_q + b_r = b_r + b_q$$

where $l_r = l_q$. But then it follows that $b_r = b_q = c$.

This procedure corresponds to the Euclidean algorithm and $l_r = l_q = t = (k, l)$ is the greatest common factor of k and l.

By running through the equations in the reverse order, one can conclude from $l_r = l_q = t$ that $k = mt$ and $l = nt$. Likewise, it follows from $b_r = b_q = c$ that $a = mc$ and $b = nc$ must hold with the same natural numbers m and n as factors.

If, for example, $l_1 > l$ and $b_2 = (m - 2n)c$ and $b = nc$ have already been found, then it follows that $b_1 = b + b_2 = (m \cdot n)c$ and $a = b + b_1 = mc$.

§10. *The Prime Numbers*

A number which can be represented as the product of two numbers different from 1 and 0 is said to be "multiplicatively composite" or just *composite*.

The numbers which are not composite, with the exception of 0 and 1, are called *prime numbers*.

If a factor of a non-vanishing product is not a natural number, then the diagram of this factor contains at least one point from which at least two arrows proceed; this will also hold for the figure of the product. It follows that: *A product of numbers is a natural number only if all factors are natural numbers. Thus the prime numbers among the natural numbers are the usual prime numbers.*

Because the multiplication of numbers causes their depths to multiply too, the following proposition holds: *Every number whose depth number is a prime number is itself a prime number.*

One can find the prime numbers most simply by deleting the numbers 0 and 1 and the composite numbers from the sequence of all numbers.

All numbers with depth number 2, 3, 5 are prime numbers. Among the numbers of depth number 4 there are composite numbers $2 \cdot 2 = 4$, $2 \cdot 2 = \{2, \{2\}\}$, $2 \cdot 2 = \{1, 3\}$ and $2 \cdot 2 = \{0, 1, 2, \{2\}\}$.

Among the numbers z such that $|z| \le 5$ there are therefore only 4 which are composite: There are $2^{65536} - 6$ prime numbers.

§11. The Decomposition into Prime Factors

If a number z is composite, then it can be expressed as the product of two numbers each of them possessing lesser depth than z itself. These factors, in as much as they are not prime numbers, can be split again into factors of lesser depth, and so on. Since the depth cannot decrease infinitely, one finally represents the number z as a product of prime factors.

If one says in addition that the number 1 is the product of 0 prime numbers, and that each prime number is the product of one prime number, then it follows that: *Every number different from* 0 *is expressable as a product of prime numbers*.

In such a decomposition the *order* of the factors is essential. In order to obtain a *unique* decomposition one would need to list the prime factors of the *natural numbers, according to magnitude.* It remains an open question, however, whether this device is sufficient to insure uniqueness.

First of all the question arises: in which cases is the product *commutative?* i. e., when does $ab = ba$ hold?

If one denotes the product of n equal factors a as the nth power of a, then the rule $a^m a^n = a^n a^m = a^{m+n}$ holds for all natural numbers m and n. Further, $mn = nm$ holds for natural numbers m and n.

It still remains to be seen, whether $ab = ba$, for $a \neq b$, is possible in other cases, for example when a and b are prime but not natural numbers. It always happens that $ab \neq ba$ when a is decomposable and b is a mononumber different from 1, since the number of knots of a is preserved in the product ab, but vanishes in ba.

There now follow a few special propositions concerning factorization: *If m is a natural number, then from $ma = mb$ it follows that $a = b$*; thus "left cancellation" holds for a natural number.

We have that $|a| = |b|$; if one represents the products ma and mb as sums of m equal terms however, it follows that $a = b$ by the lemma at the end of §7.

If $z = pa = qb$ where p and q are distinct natural prime numbers, then $z = pqc$, for some suitable c.

Since the depth $|z|$ is divisible by the distinct prime numbers p and q, and consequently also by pq; $|z|$ can be put equal to pqs. It follows that $|a| = qs$ and $|b| = ps$.

Let $p < q$. As is well known, mq can be expressed in the form $np + j$, for m = 0, 1, 2, ..., $p - 1$, and j likewise takes all the values 0, 1, 2, ..., $p - 1$. If this were not so, then two equal remainders j would have to occur, so that $np + j = mq$, and $kp + j = lq$. From this it would follow that $(k - n)p = (l - m)q$. Thus $l - m$ would be divisible by p; and if l were also allowed to assume only the values $0, 1, 2, ..., p - 1$, then it would follow that $l = m$ and $k = n$.

Now for each depth js with $j = 1, 2, 3, ..., p - 1$, the figure for b must possess a knot, since $n|b| + js$ can be put equal to $nps + js = mqs$: and the figure for z has a knot at the depth $mqs = m|a|$. Thus $b = c_1 + ... + c_p$ with $|c_j| = s$ $(j = 1, ..., p)$ and $ma = nb + c_1 + ... + c_j$.

The last term a in the sum $a + ... + a = ma$ consequently has the form $a = ... + c_1 + c_2 + ... + c_j$. Since this holds for $j = 1, 2, ..., p$ and $|c_j| = s$, so it follows by the lemma derived in §7 that $c_1 = c_2 = ... = c_p$ must hold. Should this number be put equal to c then $b = pc$, and consequently $z = qb = pqc$ and $a = qc$. With this the proposition is proved.

In general it follows that: *If $z = ma = nb$ and k is the least common multiple of the natural numbers m and n then $z = kc$ for suitable c.*

In the first instance one can take out the greatest common divisor of m and n as a factor to the left. If m and n are relatively prime, and if p is a prime factor of m, and q a prime factor of n, then by the last proposition the factor pq in ma and nb can be factored out to the left. But if, for instance, $m = 1$ so that $a = nb$, then the factor n can be taken out to the left when one replaces a by nb.

If one carries out this operation for $z = ma = nb$ as far as possible along the row, then k is finally split off entirely. So z is represented in the form $z = kc$.

It still has not been shown that in *every* prime factor decomposition of z, the product of the natural prime numbers standing on the left is always divisible by k. In every case, however, there exists a *greatest* natural number k, which can be factored out from z to the left in a *suitable representation*.

Analogous results follow for when the natural numbers occur as factors on the *right hand side*; their derivation is, however, different.

If a number z is multiplied on the right by a natural number n then each number essential in z in the diagram of z is multiplied by n, since each arrow is replaced by the figure of n.

A number essential in u is said to be a *branching number* of degree d, if it possesses d elements and $d > 1$. If z is multiplied on the right by the natural number n then the branching numbers are preserved along with their degree; they are only multiplied by n.

A row of arrows following one upon the other, and leading from the point z or from a branching number of z to another branching number or to the point 0, and which does not otherwise meet any other branching number is sais to be a *road*. The number of arrows in a road is called its *length*. If the product zn is formed, then every road aquires a length which is a multiple of n.

Now if the greatest common divisor of the lengths of all roads which occur in the figure for z is equal to m, then z can be represented in the form $z = cm$; the figure of c is obtained from the figure for z by "shortening" the roads to an mth part of their former length. The number m is the *greatest* natural number which can be factored out from z to the right; as such it is *uniquely determined*.

With this, however, it has still not been shown that in *every* prime factor decompositon of z the product of natural prime numbers standing to the right is always equal to m.

From $am = bm$ it follows, however, that $a = b$. Thus if m is a natural number right cancellation is permissible.

If $z = anb \neq 0$ where a and b are not natural numbers but n is, then one can factor a greatest natural number out of a to the right and out of b to the left and then combine these with the factor n. This yields a *greatest* natural number which can appear as a factor of z in *this position* of a decomposition.

The question presents itself as to whether things other than natural numbers can be factored out in a similar way.

§12. *The Union and Intersection of Numbers*

Since the generalized numbers are sets, the usual set theoretic operations of unions and intersections can also be applied.

The *union* $a \cup b$ of two numbers a and b is a number which possesses every element of a and every element of b and only these as elements.

The *intersection* $a \cup b$ of two numbers a and b is a number which possesses precisely those elements which occur in a and at the same time in b.

Thus e. g.

$$1 \cup 2 = 2, \quad 1 \cap 2 = 0, \quad 1 \cap 2 = 1, \quad 2 \cap 2 = 2.$$

In general,

$$0 \cup a = a \cup 0 = a; \quad 0 \cap a = a \cap 0 = 0; \quad a \cup a = a \cup a = a.$$

The operations of union and intersection are commutative and associative:

$$a \cup b = b \cup a; \ a \cap b = b \cap a;$$
$$a \cup (b \cup c) = (a \cup b) \cup c = a \cup b \cup c;$$
$$a \cap (b \cap c) = (a \cap b) \cap c = a \cap b \cap c.$$

Further the *distributive law* holds:

$$a \cup (b \cap c) = (a \cup b) \cap (a \cup c);$$
$$a \cap (b \cup c) = (a \cap b) \cup (a \cap c).$$

The following rules hold together with *addition*:

$$(a \cup b) + c = (a + b) \cup (b + c), \text{ where } a \neq 0, b \neq 0,$$
$$(a \cap b) + c = (a + b) \cap (b + c), \text{ where } a \cap b \neq 0.$$

When adding c, the 0 occuring in the figures for a, b, $a \cup b$, $a \cap b$ and which is *essential in these numbers*, is replaced throughout by the figure for c. This will yield these equalities. Since 0 is not essential in itself, however, one has to pay attention to the exceptions; e. g.

$$(0 \cup 1) + 1 = 2 \neq (0 + 1) \cup (1 + 1) = \mathbf{2} \quad \text{and}$$
$$(1 \cap 2) + 1 = 1 \neq (1 + 1) \cap (2 + 1) = 0.$$

Further we have in general:

$$a + (b \cup c) \neq (a + b) \cup (a + c),$$
$$a + (b \cap c) \neq (a + b) \cap (a + c).$$

This can easily be seen from the examples:

$$1 + (1 \cup 2) = \{\mathbf{2}\} \neq (1 + 1) \cup (1 + 2) = \{1, 2\} \quad \text{and}$$
$$1 + (1 \cap 2) = 2 \neq (1 + 1) \cap (1 + 2) = 0.$$

The following expression involving *multiplication* holds generally:

$$(a \cup b) \, c = ac \cup bc \text{ and } (a \cap b) \, c = ac \cap bc.$$

With multiplication by c all arrows in the figures for a, b, $a \cup b$, $a \cap b$ are replaced by the figure for c; the unions and intersections as such remain unchanged.

In general however:

$$a(b \cup c) \neq ab \cup ac \text{ and } a(b \cap c) \neq ab \cap ac.$$

This can be seen from the examples:

$$2 \, (1 \cup 2) = \{\mathbf{2}, \{\mathbf{2}\}\} \neq 2 \cdot 1 \cup 2 \cdot 2 = \{1, 3\} \quad \text{and}$$
$$2 \, (1 \cap 2) = 2 \neq 2 \cdot 1 \cap 2 \cdot \mathbf{2} = 0.$$

One can form the union and intersection of arbitrarily large finite collections. If z_j, $j =1, 2, ..., k$, are finitely many numbers then $\cup z_j$ signifies their *union*, that is the number which possesses precisely *all* the elements of the numbers z_j as its elements, and $\cap z_j$ signifies their *intersection*, that is the number which possesses precisely all the elements which are *common* to all the numbers z_j as its elements.

The following proposition constitutes an application: *If e_j are the elements of a number $a \neq 0$, then the product ab is equal to $\cup(b+ e_j b)$.*

For the formation of the figure ab all arrows in the figure for a had to be replaced by the figure for b; thus, in particular, those arrows leading from the point a to the elements e_j. The elements e_j themselves are then multiplied by b, that is replaced by $e_j b$. The product ab is then the union of all the numbers $b + e_j b$.

From this and the relation $0 \cdot b = 0$ there follows an *inductive definition of multiplication*: If the product ab is already defined for sets a up to a certain depth, then the proposition above yields the definition of ab for the numbers a whose depth is one greater.

First published as: "Zur Goldbachschen Vermutung", *Elemente der Mathematik* **20** (1965), 121–122 (MR **32**, 7528).

On the Goldbach Conjecture

A *natural number n* always has a *unique* predecessor; it is either $n - 1$ or 0. The word "predecessor" will be understood to mean "immediate predecessor". This is the *fundamental concept* in the theory of the natural numbers. If one takes "successor" as the basic concept and then postulates the existence of a sucessor to every natural number, one requires there to be an *infinitely* many numbers; the existence of such an infinite collection is not at all easy to prove.

Now, let *generalized numbers* (to be called simply *numbers*) have an arbitrary, *finite* number of predecessors (perhaps none). These predecessors are also generalized numbers. When "counting backwards" through these numbers one ultimately arrives at 0, a number which is without predecessors. Numbers having the same predecessors are identical.

We can obtain a directed graph for each number by taking the generalized numbers themselves as *points* and directing an *edge* from a number toward each of its immediate predecessors. The resulting graph is to contain not only the immediate predecessors of a number but its distant predecessors as well, that is the predecessors, predecessors of predecessors, etc. When the edges are directed downwards the number itself is represented as the top point of the diagram; and 0 is at the bottom.

It has been shown in Finsler [1963] that generalized numbers can be "added" and "multiplied" in a natural way by combining the associated graphs. This produces a *generalized number theory*.

The graph of the *sum* $a + b$ of two numbers a and b is obtained by "hanging" the diagram of b onto that of a so that the bottom point of a coincides with the top point of b.

The figure for the *product* $a \cdot b$ of two numbers a and b is obtained by replacing each edge in the graph of a with the graph of b where the graphs are similarly oriented. After carrying out this "substitution" one identifies points and edges that are associated with identitical numbers. This identification was not carried out in "Totally Finite Sets", Finsler [1963]. Certain alterations are therefore necessary here. For example, we have that:

$$2 \cdot 2 = \{0, 1\} \cdot 2 = \{1, 3\} = \{0, 2\} + 1.$$

Taking a number to be the *set of its predecessors*, the numbers are made to correspond to "totally finite sets". The number 0 is the empty set; the number 1 is the set whose sole element is 0, that is whose only predecessor is 0; the *natural number* 2 is the set which has 1 as its sole element. In contrast to this the *ordinal number* 2 is a set with *two* elements, 0 and 1, that is to say $2 = \{0, 1\}$.

Prime numbers are those numbers which are different from 1 which cannot be represented as the product of two other numbers which are different from 1.

One can enumerate the numbers (cf. Finsler [1963]) and one quickly finds that prime numbers dominate the list strongly: Among the first 2^{65536} numbers only six are not prime numbers.

It is remarkable that the analog of the Goldbach conjecture *fails* for the generalized numbers in spite of this abundance of primes.

Twice a number which is different from 0 and 1 is not necessarily the sum of two primes.

Doubles of the numbers $2 \cdot 2$ and $2 \cdot 2$ offer simple *counterexamples*, where **2** is the ordinal number two, $\{0, 1\}$. It can readily be seen from the diagrams for these two numbers that their representations as sums of two numbers different from 0 are the following:

$2 + 3 \cdot 2$, or $2 \cdot 2 + 2 \cdot 2$, or $3 \cdot 2 + 2$, and finally $2 \cdot 2 + 2 \cdot 2$.

None of these is a sum of two primes.

Twice the number $2 \cdot 2$ is, however, the sum of two prime numbers; it is

$$2 \cdot 2 \cdot 2 = \{0, 2\} + \{\{1, 3\}\}.$$

These two terms are of depth 2 and 5 respectively and hence are prime.

It follows that the ordinary Goldbach conjecture cannot be proven for the natural numbers using only those general principles of addition and multiplication which they share with the generalized numbers.

This counterexample is connected with the fact that there are "mononumbers" different from 1 among the generalized numbers.

Mononumbers are those numbers different from 0 which cannot be represented as the sum of two numbers different from 0.

The ordinal **2** is a mononumber, and so is any number of the form $\{0, n\}$ where n is a natural number. Thus there are infinitely many mononumbers.

Four times a mononumber a which is different from 1 is never the sum of two primes, since it has only these decompositions:

$$a + 3 \cdot a, \text{ or } 2 \cdot a + 2 \cdot a, \text{ or } 3 \cdot a + a.$$

It follows that there are *infinitely many* non-Goldbach numbers.

BIBLIOGRAPHY

Papers in this volume are marked with a *.

ACKERMANN, W. 1924 Begründung des "tertium non datur" mittels der Hilbertschen Theorie der Widerspruchsfreiheit. *Mathematische Annalen* **93**, 1–36.

– 1937 Die Widerspruchsfreiheit der allgemeinen Mengenlehre. *Mathematische Annalen* **114**, 305–315.

– 1956 Zur Axiomatik der Mengenlehre, *Mathematische Annalen* **131**, 336–345.

ACZEL, P. 1988 *Non-Well-Founded Sets*. Stanford: Center for the Study of Language and Information (CSLI lecture notes; No. 14).

ALKOR, C. 1982 Constructibility in Ackermann's Set Theory. Diss. Math. 196. Polska Akad. Nauk., Warsaw.

BAER, R. 1928a Über ein Vollständigkeitsaxiom in der Mengenlehre. *Mathematische Zeitschrift* **27**, 536–539.

– 1928b Bemerkungen zur Erwiderung von P. Finsler. *Mathematische Zeitschrift* **27**, 543.

BARWISE, J. 1988 *The Situation in Logic – IV: On the Model Theory of Common Knowledge*. Center for the Study of Language and Information, Stanford.

BARWISE, J. / ETCHEMENDY, J. 1987 *The Liar. An Essay on Truth and Circularity*. Oxford: Oxford University Press.

BARWISE, J. / MOSS, L. 1991 Hypersets. *The Mathematical Intelligencer* **13**, 31–41.

BENACERRAF, P. / PUTNAM, H. 1964 *Philosophy of Mathematics, Selected Readings*; Englewood Cliffs (N.J.): Prentice Hall. (Second revised edition: Cambridge: Cambridge University Press 1983)

BERNAYS, P. 1922 Über Hilberts Gedanken zur Grundlegung der Arithmetik. *Jahresberichte der Deutschen Mathematiker-Vereinigung* **31**, 10–19.

– 1941 Sur les questions méthodologiques actuelles de la théorie hilbertienne de la démonstration. In: Gonseth (ed.) [1941, 144–152]; Discussion, ibid. 153–161.

– 1946 (Review of Gödel [1944]) *The Journal of Symbolic Logic* **11**, 75–79.

– 1956 Zur Diskussion des Themas "Der Platonische Standpunkt in der Mathematik". *Dialectica* **10**, 262–265. Reprinted in: Finsler [1975, 148–151].

BERNSTEIN, F. 1938 The Continuum Problem. *Proceedings of the National Academy of Science* (U.S.A.) **24**, 101–104.

BIRKHOFF, G. 1937 An Extended Arithmetic. *Duke Mathematical Journal* **3**, 311–316.

– 1942 Generalized Arithmetic. *Duke Mathematical Journal* **9**, 283–302.

BOOTH, D. 1990 Hereditarily Finite Finsler Sets. *The Journal of Symbolic Logic* **55**, 700–706.

– 1991 Logical Feedback. *Studia Logica* **L, 2**, 225–239.

BREGER, H. 1992 A Restoration that Failed: Paul Finsler's Theory of Sets. In: D. Gillies (ed.), *Revolutions in Mathematics*, Oxford: Clarendon, 249–264.

BURCKHARDT, J. J. 1938 Zur Neubegründung der Mengenlehre. *Jahresberichte der Deutschen Mathematiker-Vereinigung* **48**, 146–165.

– 1939 Zur Neubegründung der Mengenlehre. Folge. *Jahresberichte der Deutschen Mathematiker-Vereinigung* **49**, 146–155.

BURALI-FORTI, C. 1897 Una questione sui numeri transfiniti. *Rendiconti del Circolo Matematico di Palermo* **11**, 154–164. (English translation: van Heijenoort [1967, 105–111].)

CARNAP, R. 1934 Die Antinomien und die Unvollständigkeit der Mathematik. *Monatshefte für Mathematik und Physik* **41**: 263–283.

CERESOLE, P. 1915 L'irréductibilité de l'intuition des probabilités et l'existence de propositions mathématiques indémonstrables. *Archives de psychologie* **15**, 255–305.

CHURCH, A. 1946 (Review of Finsler [1944]) *The Journal of Symbolic Logic* **11**, 131–132.

DAVIS, Robert L. 1943 The Number of Finite Relations. *Proceedings of the American Mathematical Society* **4**, 486–494.

DAWSON, J.W. 1984 The Reception of Gödel's Incompleteness Theorems. *Philosophy of Science Association* **2**. Reprinted in: Shanker [1988, 74–95].

DEDEKIND, R. 1918 *Was sind und was sollen die Zahlen?* Braunschweig: Vieweg (4th ed.).

FEFERMANN, S. 1988 Kurt Gödel: Conviction and Caution. In: Shanker [1988, 96–114].

*FINSLER, P. 1925 Gibt es Widersprüche in der Mathematik? *Jahresberichte der Deutschen Mathematiker-Vereinigung* **34**, 143–155 = Finsler [1975, 1–10].

*– 1926a Formale Beweise und die Entscheidbarkeit. *Mathematische Zeitschrift* **25**, 676–682 = Finsler [1975, 11–17]. (English translation with commentary: van Heijenoort [1967, 438–445].)

*– 1926b Über die Grundlegung der Mengenlehre. Erster Teil. Die Mengen und ihre Axiome. *Mathematische Zeitschrift* **25,** 683–713 = Finsler [1975, 19–49].

– 1927a Über die Grundlegung der Mathematik. *Jahresberichte der Deutschen Mathematiker–Vereinigung* **36**, 18 = Finsler [1975, 18].

*– 1927b Über die Lösung von Paradoxien. *Philosophischer Anzeiger* **2**, 183–192 = Finsler [1975, 57–66].

– 1927c Antwort auf die Entgegnung des Herrn Lipps. *Philosophischer Anzeiger* **2**, 202–203 = Finsler [1975, 69–70].

– 1928 Erwiderung auf die vorstehende Note des Herrn R. Baer. *Mathematische Zeitschrift* **27**, 540–542 = Finsler [1975, 53–55].

*– 1933 Die Existenz der Zahlenreihe und des Kontinuums. *Commentarii Mathematici Helvetici* **5**, 88–94 = Finsler [1975, 71–77].

– 1941a (Remarks in the discussion concerning Bernays [1941]). In: Gonseth (ed.) [1941, 153–161].

*– 1941b A propos de la discussion sur les fondements des mathématiques. In: Gonseth (ed.) [1941, 162–180] = Finsler [1975, 78–96].

*– 1944 Gibt es unentscheidbare Sätze? *Commentarii Mathematici Helvetici* **16**, 310–320 = Finsler [1975, 97–107].

– 1951 Eine transfinite Folge arithmetischer Operationen.
  Commentarii Mathematici Helvetici **25**, 75–90 = Finsler [1975,
  108–123].

– 1953 Über die Berechtigung infinitesimalgeometrischer
  Betrachtungen. *Convegno Internazionale di Geometria
  Differenziale*, Italia 1953, 8–12 = Finsler [1975, 124–128].

*– 1954 Die Unendlichkeit der Zahlenreihe. *Elemente der
  Mathematik* **9**, 29–35 = Finsler [1975, 129–135].

*– 1956a Der platonische Standpunkt in der Mathematik. *Dialectica*
  10, 250–255 = Finsler [1975, 136–141].

*– 1956b Und doch Platonismus. *Dialectica* **10**, 266–270 = Finsler
  [1975, 152–156].

– 1956c Briefwechsel zwischen P. Lorenzen und P. Finsler.
  Dialectica **10**, 271–277 = Finsler [1975, 157–163].

*– 1963 Totalendliche Mengen. *Vierteljahresschrift der
  Naturforschenden Gesellschaft in Zürich* **108**, 141–152 = Finsler
  [1975, 164–175].

*– 1964 Über die Grundlegung der Mengenlehre. Zweiter Teil.
  Verteidigung. *Commentarii Mathematici Helvetici* **38**, 172–218 =
  Finsler [1975, 176–222].

*– 1965 Zur Goldbachschen Vermutung. *Elemente der Mathematik*
  20, 121–122 = Finsler [1975, 223–224].

– 1969 Über die Unabhängigkeit der Kontinuumshypothese.
  Dialectica **23**, 67–78.

– 1975 *Aufsätze zur Mengenlehre* (ed. by G. Unger). Darmstadt:
  Wissenschaftliche Buchgesellschaft.

FRAENKEL, A. 1922a Zu den Grundlagen der Cantor–Zermeloschen
  Mengenlehre. *Mathematische Annalen* **86**, 230–237.

– 1922b Der Begriff "definit" und die Unabhängigkeit des
  Auswahlaxioms. *Sitzungsberichte der Preussischen Akademie der
  Wissenschaften* (1922), 253–257 (English translation: van
  Heijenoort [1967, 284–289]).

– 1923 *Einleitung in die Mengenlehre*. Berlin: Springer (2nd ed.)

– 1925 Untersuchungen über die Grundlagen der Mengenlehre.
  Mathematische Zeitschrift **22**, 250–273.

- 1926 Die Gleichheitsbeziehung in der Mengenlehre. *Journal für die reine und angewandte Mathematik* **157**, 79–81.

- 1928a (Review of Baer [1928a]) *Jahrbuch Fortschritte der Mathematik* **54**, 90.

- 1928b (Review of Finsler [1928]) *Jahrbuch Fortschritte der Mathematik* **54**, 90.

- 1928c (Review of Baer [1928b]) *Jahrbuch Fortschritte der Mathematik* **54**, 90.

- 1928c *Einleitung in die Mengenlehre*. Berlin: Springer (3rd ed.)

FRAENKEL, A. / BAR-HILLEL, Y. 1958 *Foundations of set theory*. Amsterdam/New York/Oxford: North Holland.

FRAENKEL, A. / BAR-HILLEL, Y. / LEVY, A. 1973 *Foundations of set theory*. Amsterdam/New York/Oxford: North Holland (Second revised edition).

FREGE, G. 1884 *Die Grundlagen der Arithmetik*. Breslau.

- 1893 *Grundgesetze der Arithmetik*, Vol. I. Jena.

- 1903 *Grundgesetze der Arithmetik*, Vol. II. Jena.

GEULINCX, A. 1663 *Methodus inveniendi argumenta*. In: Arnoldi Geulincx Antverpiensis Opera Philosophica, rec. J.P.N. Land, Vol. II (1892), p. 25.

GODDARD, L. / JOHNSTON, M. 1983 The Nature of Reflexive Paradoxes: Part I, *Notre Dame Journal of Formal Logic* **24**: 491–508.

GÖDEL, K. 1931 Über formal unentscheidbare Sätze der Principia Mathematica und verwandter Systeme I. *Monatshefte für Mathematik und Physik* **37**, 173–198.

- 1938 The consistency of the axiom of choice and of the generalized continuum–hypothesis. *Proceedings of the National Academy of Sciences* (U.S.A.) **24**, 556–557.

- 1944 Russell's Mathematical Logic. In: P. Schilpp (ed.), *The Philosophy of Bertrand Russell* Evanston-Chicago: Northwestern University Press, 123–153. Reprinted in: Benacerraf/Putnam [1964, 211–232; 1983, 447–469].

- 1964 What is Cantor's continuum problem? In: Benacerraf/Putnam [1964, 258–273; 1983, 470–485].

GONSETH, F. (ed.) 1941 *Les entretiens de Zurich sur les fondements et la méthode des sciences mathématiques, 6–9 décembre 1938.* Zurich: Leemann frères & Cie.

GRELLING, K. / NELSON, L. 1908 Bemerkungen zu den Paradoxien von Russell und Burali-Forti. *Abhandlungen der Fries'schen Schule*, Neue Folge, Band II, Heft **3**: 301–324. In: L. NELSON, *Beiträge zur Philosophie der Logik und Mathematik.* Frankfurt am Main: Verlag Öffentliches Leben 1959, pp. 59–77.

GREWE, R. 1969 Natural Models of Ackermann's Set Theory. *Journal of Symbolic Logic* **34**, 481–488.

GROSS, H. 1971 Nachruf: Paul Finsler [with complete bibliography of Finsler's papers]. *Elemente der Mathematik* **26**, 19–21 = Finsler [1975; IX–X, 241–242].

– 1975 Geleitwort [Preface to:] Finsler [1975, VII–VIII].

GUT, B. 1979 *Inhaltliches Denken und formale Systeme.* Oberwil: Verlag Rolf Kugler.

HEIJENOORT, J. VAN (ed.) 1967 *From Frege to Gödel – A source book in mathematical logic 1879–1931.* Cambridge (Mass.): Harvard University Press.

HILBERT, D. 1900 Mathematische Probleme. Vortrag, gehalten auf dem internationalen Mathematiker-Kongress zu Paris 1900. *Nachrichten von der Königlichen Gesellschaft der Wissenschaften zu Göttingen* (1900), 253–297 = Gesammelte Abhandlungen, Vol. 3, 290–329.

– 1913 *Grundlagen der Geometrie.* Leipzig/Berlin: Teubner (4th ed.). Appendix VI: Über den Zahlbegriff.

– 1922 Neubegründung der Mathematik. (Erste Mitteilung). *Abhandlungen aus dem Mathematischen Seminar der Hamburgischen Universität* **1**, 157–177.

– 1923 Die logischen Grundlagen der Mathematik. *Mathematische Annalen* **88**, 151–165.

– 1926 Über das Unendliche. *Mathematische Annalen* **95**, 161–190. (English translation: van Heijenoort [1967, 367–392]).

HILBERT, D. / BERNAYS, P. 1934/39 *Grundlagen der Mathematik.* Berlin: Springer, Bd. I: 1934, Bd. II: 1939.

HINTIKKA, K. J. 1957 Vicious Circle Principle and the Paradoxes. *Journal of Symbolic Logic* **22**, 245–249.

HÖSLE, V. 1986 Die Transzendentalpragmatik als Fichteanismus der Intersubjektivität. *Zeitschrift für philosophische Forschung* **40**: 235–252.

HUGHES, G. E. 1982 *John Buridan on Self-reference*. Cambridge: Cambridge University Press.

KESSELRING, T. 1984 *Die Produktivität der Antinomie*. Frankfurt am Main: Suhrkamp.

KLEENE, S.C. / ROSSER, J.B. 1935 The inconsistency of certain formal logics. *Annals of Mathematics* **36**, 630–636.

KREISEL, G. 1954 (Review of Finsler [1954]). *Mathematical Reviews* **15**, 670.

LAKE, J. 1975 Natural Models and Ackermann-Type Set Theories. *Journal of Symbolic Logic* **40**, 151–158.

LEVY, A. 1959 On Ackermann's Set Theory. *Journal of Symbolic Logic* **24**, 154–166.

LIPPS, H. 1923 Die Paradoxien der Mengenlehre. *Jahrbuch für Philosophie und Phänomenologische Forschung* **6**, 561–571.

– 1927 Entgegnung. *Philosophischer Anzeiger* **2**, 193–201.

LORENZEN, P. 1955 (Review of Finsler [1954]). *Zentralblatt der Mathematik* **55**, 46.

– 1956 Briefwechsel zwischen P. Lorenzen und P. Finsler. *Dialectica* **10**, 271–277. Reprinted in: Finsler [1975, 157–163].

MAZZOLA, G. 1969 Finslersche Zahlen. *Commentarii Mathematici Helvetici* **44**, 495–501.

– 1972 Der Satz von der Zerlegung Finslerscher Zahlen in Primfaktoren. *Mathematische Annalen* **195**, 227–244.

– 1973 Diophantische Gleichungen und die universelle Eigenschaft Finslerscher Zahlen. *Mathematische Annalen* **202**, 137–148.

MIRIMANOFF, D. 1917a Les antinomies de Russell et de Burali-Forti et le problème fondamental de la théorie des ensembles. *L'enseignement Mathématique* **19**, 37–52.

– 1917b Remarques sur la théorie des ensembles et les antinomies Cantorienns, – I. *L'enseignement Mathématique* **19**, 209–217.

– 1920 Remarques sur la théorie des ensembles et les antinomies Cantorienns, – II. *L'enseignement Mathématique* **21**, 29–52.

NELSON, E. 1986 *Predicative Arithmetic*. Princeton: Princeton University Press.

NEUMANN, J. VON 1925 Eine Axiomatisierung der Mengenlehre. *Journal für die reine und angewandte Mathematik* **154**, 219–240 und **155**, 128 (English translation: van Heijenoort [1967, 393–413]).

– 1928 Die Axiomatisierung der Mengenlehre. *Mathematische Zeitschrift* **27**, 669–752.

QUINE, W. V. 1962 Paradox. *Scientific American* **206**: 84–96.

RAMSEY, F. P. 1926 The foundations of mathematics. *London Mathematical Society, Proceedings*, 2nd Series **25**: 338–384.

REINHARDT, W. 1970 Ackermann's Set Theory Equals ZF. *Annals of Mathematical Logic* **2**, 149–249.

RICHARD, J. 1905 Les principes des mathématiques et la problème des ensembles. *Revue générale des sciences pures et appliquées* **16**, 541.

RÖSCHERT, G. 1985 *Ethik und Mathematik*. Stuttgart: Freies Geistesleben.

RUSSELL, B. 1908 Mathematical logic as based on the theory of types. *American Journal of Mathematics* **30**: 222–262.

SCOTT, D. 1960 *A Different Kind of Model for Set Theory*. Unpublished paper read at the International Congress of Logic and Method, Stanford.

SHANKER, S.G. (ed.) 1988 *Gödel's Theorem in Focus*. London- New York-Sydney: Croom Helm.

SKOLEM, T. 1922 Einige Bemerkungen zur axiomatischen Begründung der Mengenlehre. *Proceedings of the 5th Scandinavian Math. Congr.*, Helsinki 1922, 217–232 (English translation: van Heijenoort [1967, 290–301]).

– 1926 (Review of Finsler [1926b]) *Fortschritte der Mathematik* **52**, 192–193.

SPECKER, E. 1954 Die Antinomien der Mengenlehre. *Dialectica* **8**, 234–244.

TARSKI, A. 1935 Der Wahrheitsbegriff in den formalisierten Sprachen, *Studia Philosophica* (1935), 261–405.

UNGER, G. 1975a Vorwort des Herausgebers zu Paul Finslers *Aufsätze zur Mengenlehre*. In: Finsler [1975, XI–XVI].

– 1975b Referat der Arbeit über ein Vollständigkeitsaxiom in der Mengenlehre von Reinhold Baer, Freiburg. In: Finsler [1975, 50–52].

– 1989 *Die Rettung des Denkens*. Stuttgart: Freies Geistesleben (2nd ed.).

VARDY, P. 1979 Some remarks on the relationship between Russell's Vicious-Circle Principle and Russell's paradox. *Dialectica* **33**, 3–19.

VIELER, H. 1926 *Untersuchungen über die Unabhängigkeit und Tragweite der Axiome der Mengenlehre usw.*, Diss. Marburg.

WANDSCHNEIDER, D. 1993 *Das Antinomienproblem und seine pragmatische Dimension*. In: H. Stachowiak (ed.): *Pragmatik: Handbuch des pragmatischen Denkens*, Vol. IV: *Sprachphilosophie, Sprachpragmatik und formative Pragmatik* (Hamburg: Meiner 1993), p. 320–352.

WEBB, J.C. 1980 *Mechanism, Mentalism, and Metamathematics. An Essay in Finitism*. Dordrecht: Reidel.

WEYL, H. 1946 Mathematics and Logic. *The American Mathematical Monthly* **53**, 2–13.

WITTENBERG, A. 1953 Über adäquate Problemstellungen in der mathematischen Grundlagenforschung. *Dialectica* **7**, 232–254.

– 1956 Warum kein Platonismus? Eine Antwort an Herrn Prof. Finsler. *Dialectica* **10**, 256–261. Reprinted in: Finsler [1975, 142–147].

WITTENBERG, A. et al. 1954 (Discussion of Wittenberg [1953]). *Dialectica* **8**, 145–157.

ZERMELO, E. 1908 Untersuchungen über die Grundlagen der Mengenlehre I. *Mathematische Annalen* **65**, 261–281 (English translation: van Heijenoort [1967, 199–215]).

ZIEGLER, R. 1995 *Selbstreflexion. Studien zur Selbstbeziehbarkeit im Denken und Erkennen*. Dornach: Verlag am Goetheanum.

Index

Absolute
- consistency   12ff., 61ff., 72, 141, 154, 225f.
- decidability   58ff.
- logic   103, 124, 226
- truth   19, 180
- undecidability   55, 59ff., 63
Accessibility   249, 251
ACKERMANN, Wilhelm   34f., 71, 77, 99ff.
ACKERMANN Set Theory   77
ACZEL, Peter   3, 83, 86f., 245, 250f., 254ff., 270f.
Adjacency matrix   252f.
ALKOR, C.   100
Allmenge   31f., 99ff., 103, 122, 200ff.
Anti-Foundation Axiom   245, 255f.
Antinomies, *see* Paradox
ARCHIMEDEAN order   208ff., 212f.
Arithmetic
  consistency of -   24, 33
  incompleteness of - 13, 16, 53ff.
Axiom I, II, III, *see* FINSLER Set Theory
Axiom of Choice, *see* Choice, Axiom of
Axiom of Foundation, *see* Foundation, Axiom of
Axiomatic method   24
Autological, *see also* Paradox of GRELLING-NELSON   44ff.

β, *see* Beta relation
BAER, Reinhold   89ff., 141, 156, 193ff., 211ff., 219
BARWISE, Jon   88
BERNAYS, Paul   3, 34, 54, 58, 67f., 76f., 96, 101, 104,
                        109, 154, 160, 214, 218
BERNSTEIN, Felix   87ff.
Beta relation   81, 105, 113, 140, 168, 172, 194ff.,
                        207f.
Binomial coefficients   259ff.
BIRKHOFF, Garrett   244
BOLZANO, Bernhard   166
BROUWER, Luitzen E. J. 23, 102
BURALI-FORTI paradox, *see* Paradox
BURCKHARDT, Johann J.   162
BURNSIDE's theorem   268

CHRONOLOGY OF PAUL FINSLER'S LIFE

11. April 1894	Born in Heilbronn (Neckar), Germany. Grammar-school in Urach.
1908–12	Secondary school in Cannstatt with scientific emphasis.
1912–13	Studies at the Technische Hochschule in Stuttgart, with Rudolf Mehmke and Wilhelm Kutta among others.
1913–18	Graduate studies in mathematics in Göttingen with Erich Hecke, David Hilbert, Felix Klein, Edmund Landau, Carl Runge, Ludwig Prandtl, Max Born and Constantin Carathéodory among others.
1918	Promotion to Ph. D. through C. Carathéodory. Dissertation: "Über Kurven und Flächen in allgemeinen Räumen" (Curves and Surfaces in General Spaces).
1922	Appointment as a university lecturer ("Habilitation") at the University of Köln.
1923	Inaugural lecture at the University of Köln: "Gibt es Widersprüche in der Mathematik?" (Are there Contradictions in Mathematics?).
1924	Discovery of a comet.
1926	The paper "Formale Beweise und Entscheidbarkeit" (Formal Proof and Decidability) anticipates the conceptual (not formal) core of Kurt Gödels first incompleteness theorem from 1931. "Über die Grundlegung der Mengenlehre. Erster Teil: Die Mengen und ihre Axiome" (On the Foundations of Set Theory, First part: Sets and their Axioms).
1927	Associate professor for applied mathematics, in particular descriptive geometry at the university of Zürich. Research topics: geometry, in particular differential geometry; elementary number theory; probability theory; foundations of mathematics, in particular set theory. Regular one-year classes with exercises, four hours per week, in descriptive geometry.

1934	With Elie Cartan's little book "Les espaces de Finsler" (Paris: Hermann & Cie 1934) the name "Finsler space" becomes common-place in differential geometry.
	Finsler spaces or *Finsler manifolds* are generalisations of Riemannian spaces, where the general definition of the length of a vector is not necessarily given in the form of the square root of a quadratic form and where the Minkowskian geometry holds locally.
1937	Discovery of a comet.
1944	Full professor in Zürich after the leave of Andreas Speiser to the University of Basel. Regular introductory classes in calculus; beginning in fall 1946 descriptive geometry again.
1951	Reprint of the dissertation by Birkhäuser Verlag (Basel) with extensive bibliography up to 1949 by H. Schubert (Lehrbücher und Monographien aus dem Gebiete der exakten Wissenschaften – Mathematische Reihe, Band 11).
1958	"Vom Leben nach dem Tode" (On Life after Death). Published by an association of scientists on behalf of the orphanage in Zürich.
1959	Retires from active teaching duties and becomes honorary professor.
29. April 1970	Death in Zürich on his way to the Dies Academicus at the University of Zürich.

Sources:

Peter Stadler (ed.) *Die Universität Zürich 1933–1983. Festschrift zur 150-Jahr-Feier der Universität Zürich* (Published by the Rektorat of the University of Zürich). Zürich 1983.
Johann Jakob Burckhardt: *Die Mathematik an der Universität Zürich 1916–1950 unter den Professoren R. Fueter, A. Speiser und P. Finsler.* Basel: Birkhäuser 1980 (Elemente der Mathematik – Beiheft Nr. 16). [Contains more sources and a complete bibliography of Finsler's papers.]
Herbert Gross: "Nachruf Paul Finsler". *Elemente der Mathematik*, Vol. 26, 1971, pp. 19–21.

Erratum

David Booth / Renatus Ziegler (ed.): *Finsler Set Theory: Platonism and Circularity*. Basel: Birkhäuser Verlag 1996.

CORRIGENDA:
Replace the index on pp. 271–276 with the following new index: